The Weaponization of Weather in the Phony Climate War

The Weaponization of Weather in the Phony Climate War

JOE BASTARDI

EDITED BY JORDAN PAYNE

FOREWORD BY SEAN HANNITY

gatekeeper press

Columbus, Ohio

The Weaponization of Weather in the Phony Climate War

Published by Gatekeeper Press
2167 Stringtown Rd, Suite 109
Columbus, OH 43123-2989
www.GatekeeperPress.com

ISBN (paperback): 9781662903656
eISBN: 9781662903663

Table of Contents

Foreword

There is an excellent chapter in Joe Bastardi's first book, *The Climate Chronicles: Inconvenient Revelations You Won't Hear From Al Gore — And Others,* that has gained traction over the years. It's called "The Weaponization of Weather." I advised Joe to write an entire book on it. He didn't just take up the challenge; he produced a tour de force.

Joe's 45 years of experience in the field of meteorology is evident in this book, as is his impressive knowledge of history. *The Weaponization of Weather in the Phony Climate War* details past meteorological events that were even more severe than what we've seen in more recent times. It also exposes the ways in which climate alarmists are weaponizing floods, droughts, tornadoes, warm weather, cold weather, hurricanes — heck, even the coronavirus. Joe offers facts that show it's almost exclusively natural variability that runs the climate, not man.

Joe has rightfully come to the conclusion that this "phony war" is a smoke screen to push another agenda — the destruction of the very system that made this nation successful. We fight not over the climate but to preserve life, liberty, and the pursuit of happiness. We fight not because we are unconcerned about the environment but because we believe a top-down system of government is deleterious. The battle isn't over climate change. It's over our very way of life.

Joe and I agree: Never in our wildest dreams did we think the climate-change debate would devolve into the cesspool that has become politics today.

Our youth have been indoctrinated into believing they have no future without draconian measures to stave off climate change. An entire chapter of this book is devoted to showing them that they *do* have a future. Our youth are becoming a pivotal voting bloc. And as Joe aptly points out, we have to address them. We can achieve that by de-weaponizing the issue with free-market-based energy ideas.

Joe's objective in this book is to make CO_2 a moot point by using his opponents' own arguments against them. But he also offers up countermeasures that enhance, not destroy, our freedoms. They are real solutions, not phony smoke screens. If you're concerned about CO_2, Joe provides you an opportunity to do something about it — but in a way that doesn't take trillions of dollars out of the economy or shackle our freedoms.

President Ronald Reagan famously said, "Freedom is a fragile thing and is never more than one generation away from extinction. It is not ours by inheritance; it must be fought for and defended constantly by each generation, for it comes only once to a people. Those who have known freedom and then lost it have never known it again." The next generation is here, and climate alarmism is all about manipulating them into voting for statists. Joe understands, like Reagan did, that freedom lost is never known again. It is my hope that the reader will conclude the same thing after reading this masterpiece. Let's do our part to ensure that this harmful smoke screen gets thoroughly doused.

Sean Hannity
Author, Syndicated Radio Talk-Show Host, and Fox News Anchor

Acknowledgements

First of all, to my Heavenly Father, from whom all blessings flow. Those blessings take the form of my family, my friends, and the many great lights that have come before me.

Thanks to my editor, Jordan Payne, who takes what I send him and turns it into a solid expression of the points I am trying to get across. Given what the rough drafts look like, it borders on miraculous.

The Penn State wrestling staff has been a big inspiration to me. Never have I seen people with so much talent with so little ego. Had I not wrestled, I wouldn't have gone as far as I did. Those tests became my lessons.

There are numerous rocks I lean on. You know who you are.

A huge shout-out to Sean Hannity, who challenged me to write this book when I was on set with him last year.

And finally, thanks to the great people on both sides of the issue. That's right. Even the scientists with whom I disagree are brilliant. Sadly, many outside distractions have developed. But the amount of work, research, and advancement in this field are amazing. If this kerfuffle boiled down to simply the exchanging of ideas, the political sewer this has become wouldn't exist.

Hopefully, we can stem this tide and let freedom ring.

Photography: vjmStudios.com
Representation: MastheadGroupLLC.com
Book Cover Design: Chuck Miller, ChuckMillerMedia.com

Introduction

The weaponization of weather is a tactic I saw emerging years ago. At the time, I predicted climate-change doomsayers would ramp up its application. They did, so I will now flesh out their machinations.

As someone who has loved weather and climate from childhood, what's particularly distressing to me are the militaristic terms being used, which is part of an effort to elevate the status of the "combatants" so they feel they are part of some movement greater than themselves.

I believe altruism is instilled in everyone, and that's great. The problem is that this yearning is being exploited beyond recognition. This pains those of us who believe past weather events provide valuable clues for what lies ahead. By exaggerating current events, climate-change doomsayers are doing a disservice to the bygone *true heroes* of this nation. In my opinion, a form of self-worship has developed. Relatively speaking, our lives have never been better. And yet people are being coaxed into action.

The "weapon" in this age of instant communication is depicting *all* events as out of the ordinary. It works, because many people have a poor or even nonexistent understanding of the past. This weapon is a form of soft tyranny — for now. Given the anger that's developing, this soft tyranny could soon turn unyielding.

I have made no secret about my approach to global warming, which is simply questioning the amount of impact CO_2 has on the planet. I won't rehash it all here (see my first book for more information on greenhouse gases), but I will briefly explain why I'm an objective broker.

- It does me no good when the weather is tranquil. Has anyone ever seen me on TV when pleasant conditions are engulfing the U.S.? How can I promote my work by declaring, "Wow! The weather is so nice and perfect"? The fact is, bad weather drives business. So why would I *not* be on "the weather is getting worse and you need my help to navigate through it" bandwagon?
- Suppose it is warming. So what? Some arguments indicate that's a good thing. For example, anomalously warm epochs in the past are known as climate *optimums*. Ergo, what's the big deal? We should do what we've always done — face the challenge and adapt to it. That's what a free and competitive society would do.
- When it comes to the cause of the warming, it really doesn't matter. I still have to deal with it. If tomorrow I conclusively decide I was wrong on the attribution, it makes no difference. The man on the moon could be sprinkling magic dust, but I still have to deal with it.

You can't win in life until you are no longer afraid to lose. But how can you surrender to the reality that you may be wrong? One of the big problems I see in the climate debate is that many of the people clamoring about the warming have never had to make a forecast. Which means they are unable to balance all the factors when it comes to nailing down the cause. It's the total picture that determines the outcome, and one major sticking point is that CO_2 has *never* been the climate control knob when considering the entire geological history of the earth. That's why many in the geological community have rejected the CO_2 narrative. Geologists by and large remain silent for their own good.

This book examines and compares past events with the hope that a rational person will at least take a look. There's a lot of anger out there. I see it all the time on social media. As for me, my real motive is to get closer to God by responding to the challenges in this field. God never gives us the answers. He tells us to utilize the individual gifts He provided to more clearly see His presence. You only advance through being challenged, for unto all good there must be opposition. Otherwise, how would you know what's good?

You may not agree, and that's fine. But I've learned that God always tests us beyond our capability so that we may come to realize the only way to make headway is to learn and, as Robert Browning said, reach beyond your grasp.

Overall, I'm a happy person. I can't save the planet, and I am a hero to no one. But every day I get up is like Christmas, and the gift I get to unwrap is the majesty and challenge of forecasting. It's sad the weather dialogue has become what it has, but that's the way the devil works. He tries to hide the majestic evidence of a loving Heavenly Father with anything he can.

Perhaps the real war is on a level even higher than what I outlined here — a spiritual one pitting relative truth against absolute truth. But that's another subject for another time.

I use the word phony in this book for two reasons.

First, if someone is going to label a reasoned exchange of ideas as a war, that "war" consists only of shutting down the exchange of ideas. Which is what this nation has gone to war over — to fight those who sought to deny basic freedoms by means of control. Most of the people involved in this (not all, but most) have never seen combat. How would they know what real war is? They can't know. Yet they call this debate a war.

Unless you have been in combat (I have not), you have no idea what it's like, even if you were told stories by shell-shocked veterans. My wrestling coach at Penn State was one of the first men on the beaches at Normandy. He's the reason I'm dressed the way I am on the front cover. Admittedly, it's an absurd look: A water pistol machine gun with flowers sticking out of it, green camouflage, American flag bandana… It's the look of a fake warrior. I am no warrior, to be sure, nor am I in a war. But I may as well dress like it since the other side is deeming this warfare.

Second, climate change is now a political issue. The so-called science is a smoke screen. This book will reveal how what you are being told is designed to cause fear and to push people to trade away their freedoms. Like Don Quixote fighting windmills, there is no true hope for a solution. Climate doomsayers want only to utilize this debate to influence the people around them. The key for true freedom lovers is to offer solutions that will take off the political edge. Take the issue out of their hands in a way that's not an economic suicide pact. The last chapter of the book will explain that.

To me, this is a phony war. Admittedly, I like using satire to demonstrate a point, but it's insulting to those who have put their lives on the line to conflate climate change and war. In the end, the basic question that needs to be answered is this: Will you sacrifice life, liberty, and the pursuit of happiness for the supposed safety being offered by people who are heavily invested in political expediency?

The words from Andrea Bocelli's love song to God, "Un Nuovo Giorno," ring true: "Solo rischiando tu vivrai" *("Without risking there is no life").* Combine that with Ronald Reagan's admonition that it only takes one generation to lose freedom and Dwight D. Eisenhower's farewell speech on the dangers of outsourcing the pursuit of knowledge to government and academia. Interestingly enough, the left never acknowledges that part of the speech. Eisenhower wasn't anti-science; he was pro-freedom. From the speech:

> *Akin to, and largely responsible for the sweeping changes in our industrial-military posture, has been the technological revolution during recent decades. In this revolution, research has become central, it also becomes more formalized, complex, and costly. A steadily increasing share is conducted for, by, or at the direction of, the Federal government. … The prospect of domination of the nation's scholars by Federal employment, project allocation, and the power of money is ever present and is gravely to be regarded. Yet in holding scientific discovery in respect, as we should, we must also be alert to the equal and opposite danger that public policy could itself become the captive of a scientific-technological elite.*

This is not a war. And portraying it as such should raise questions.

"The heavens declare the glory of God; the skies proclaim the work of his hands."—Psalm 19:1

CHAPTER 1

What's This Really About?

Where you stand today was built yesterday to reach for tomorrow. My previous book touched only slightly on this in the chapter entitled, "In God We Trust — or Do We?"

Think about your own family history. You got to where you are now with hard work and determination, but people before you set the stage. If you're like me, then you see a synergism between your ancestors and a mighty and merciful God who afforded us a chance to even have a life. You can disagree with that, but what you can't disagree with is that there was no preordained human reason for why you were born. I never begrudge another man's beliefs, but if you think creation was indiscriminate, then your faith system is one that puts randomness in control.

Life, liberty, and the pursuit of happiness had nothing originally to do with you. The Founding Fathers discerned this. They understood not just gratitude but the idea that we were all born with free will. However, each of us is also given certain talents that, if fully developed, can elevate us and make life worth living. The purpose of our system of government was to afford men the chance to excel by cultivating gratitude for both those who came before us and our Heavenly Father who set it up in the first place. There is natural synergism between the past, the present, and the future. I'm more concerned about where I stand in front of my Heavenly Father. Am I doing what I was *made* to do?

It comes down to the relative truth of man versus the absolute truth of God.

If you don't believe in God, or if you believe in man more than God, you are naturally going to try to replace absolute truth with what you think is best. The book *The Half-Life of Facts* shows that knowledge is expanding so rapidly, we can hardly keep up with it. If something is demonstrably false, that means it was never true. Man may have *thought* it to be true, but in reality, it's proportional to what they knew and saw at the time.

What's happening today is that, as human knowledge increases, so does arrogance, which becomes a quest to *be* God. But God's truth is always there. It's not going to change, no matter what man may discover. Science and God go hand in hand — and have to. Otherwise, we're left with incomprehensible random phenomena.

If you desired to fundamentally change the nation — a nation built on principles that pay homage to a mighty and merciful God — what must you do to vanquish it? You must destroy its foundation. You would find out every little thing the nation did wrong and proclaim it as loudly as possible. You would do it not to unite but to divide. But you would also give yourself a pass, assuming you — in your infinite wisdom — would never have made the same mistakes or thought the same way.

This is pure arrogance. And the Bible teaches us that pride, which is a form of arrogance, is the greatest sin.

Speaking of sin, a devious-minded revolutionary would also seek to destroy the belief system of traditionalists. This tactic — lies, deception, caterwauling — is as old as man himself, and it's always used to obscure what is righteous. In the end, there's no foundation upon which to stand.

If you desired to destroy this nation's foundation, are lies, deception, and caterwauling not the kind of subterfuge you would use? Would you not disguise the agenda by claiming it's geared toward improving the nation? And would you not try to discredit all the good that came hitherto?

You may say, "Why would a loving and merciful God allow such things?" Think about it. If you could beat the devil on your own, why would you need God? If God wants you to draw closer to Him, does it not stand to reason that tests will be placed before you to make you turn toward Him?

My mission isn't to save the world. It's to nail forecasts and to show *how* through the meticulous study of the past. I believe that God's greatest gift to man is free will — the chance to take the test offered in life. Or, to borrow from Robert Browning again, reaching beyond your grasp. You can't reach beyond your grasp if you are shackled by men who want to replace God with a One World Government. If government replaces God, then from whom (or what) are you forced to seek guidance?

Truly driven people like me are searching for the truth behind the cause of global warming — which, by the way, is approaching what was always referred to as a climate *optimum*, not a climate emergency.

The simple, undeniable fact is that life is better when it's warmer. Yet the fear is that today's warming will simply keep feeding back, turning the earth into a hellhole. The problem is, there are natural limits to that feedback, and there is no actual proof CO_2 is the primary catalyst. Yet not a single doomsayer will point this out or give the time of day to Dr. William Happer or Dr. Roy Spencer or Dr. Willie Soon or the many other science giants who question CO_2's effects.

What really angered me was the trashing of Dr. Gray, who pushed the science of hurricane forecasting to levels not imagined in the 1970s. During a tame hurricane period, Dr. Gray said he expected to see a $250 billion storm. He has since passed away, but eventually he will be right, and it will have nothing to do with CO_2. It will be due to the erection of huge targets.

As a wrestler, I learned that you must know and understand—but not fear—your opponent. Thus, I naturally look more at what people who disagree with me are saying and analyze their opinions. As a forecaster, I know there's a chance my predictions may be wrong. But when you desire credit merely for your *ideas*, then what you own owns you. Because I set up weather events beforehand, I get married to them, and so occasionally my prognostications get trounced. But only the scars of defeat lead to future triumph.

We all must be open-minded. Unfortunately, actors and activists are anything but undogmatic. Sadly, that includes the media. Former Ohio Governor John Kasich, back when I appeared on his show "Heartland," would talk to me regularly off the air. He once warned me to never get seduced by the media. I often think about that remark. I have an ego (I like being on television), but not at the price of who I am and what I was made to do. The balance between getting warranted attention to get your message out and craving that attention is a challenge.

If, as a scientist, you are hanging out with the big shots, it's hard to let go. I don't know about some of these other guys, but I was a school nerd and began wrestling and weightlifting because I couldn't stand being bullied on the playground. Short fat kids with funny-sounding last names were bullying targets. Interestingly, I was bullied after I began living in New Jersey as a 5th grader. But there was no sign of it in grades 1-4 in Texas, where God was very much present in our classroom. In any event, my early life experience ridded me of the attention cravings to some extent.

I am who I am, and what you see is what you get. If you see me on TV, that's how I talk and act all the time. I'm not an actor, and, as one of my friends told me, I have a face for radio.

The bottom line? For many people, there's an intersection, if you will, of second-degree ignorance and good intentions. But the road to hell is paved with good intentions.

One might say, "Well, Joe, you're the same way — only the opposite." That's where the *mission* comes into play. I am not here to save the world, and that belief is buttressed by my faith. From my perspective, the planet needs to take action on obvious issues, not a cryptic one like climate change.

One such urgent issue is homelessness. I was in Washington, DC, recently, and during the visit I wondered how people who live and work there can genuinely talk about climate change as being a pressing problem when there are people living in tents under nearby bridges. Nobody is going to remember me when I'm gone. Personally, I couldn't care less. But if your seductive objective is to be remembered as the guy who saved the world, it's easy to overlook the problems staring you in the face.

As for politicians, many of them know darn well they can easily influence the younger demographics that have been subjected to a constant bombardment of climate-alarmist dogma. Ask them what the average temperature of the planet throughout its history is. Ask them what the average amount of CO_2 is. Ask them what they want the average temperature and amount of CO_2 to be — and why. Ask them if they've ever looked at the saturation mixing ratio and how it relates to dew point temperatures. Yet all they hear about are things like a Green New Deal costing trillions of dollars that would theoretically stave off only .01°C over 30 years.

Here's my question to them: Are you willing to go to war with China because it builds a coal plant *every week*? After all, if you're fussing at our fossil-fuel industry (which, to its credit, has been reducing emissions), why are you tolerating what's going on outside the U.S.? Particularly when you're so One World oriented? It makes no sense. But it's a perfect example of an education system that's moving toward indoctrination by authority rather than the questioning of authority.

Ironically, the challenge I posed above would be expected in the 1970s, when I was a liberal and taught to dig deeper. I think the term "liberal" is now a misnomer, and purposely so.

Many prominent people are using downright scary rhetoric that says we have ruined our kids' future. Yet they enjoy luxuries that none of us could possibly envision without the progress made possible by fossil fuels. Without them, the world would be a far harsher place for all of us. Go back and look at how life used to be.

It's all about One World governance and the redistribution of wealth. Not too many things grow top-down, rather they grow bottom-up, which provides a hint as to the natural design of adaptation and progress — something that these alarmists are obviously against.

Finally, you don't have to believe in God in order for me to see His light in you. I don't judge someone because I disagree on a particular issue. My faith exposes my own numerous weaknesses. There are times I fall very short. But as a kindly old priest taught me, God is the search for *who*; science is the search for *why*. (Suggested reading: *Modern Physics and Ancient Faith* by Dr. Stephen Barr.) The priest said find the who and you will see the why. See the why and it will point back at who. Which in essence is what Barr's book is about.

There is no fight between God and science, only barriers constructed by men who think they are smarter than God. By believing we can control the climate, we are portraying ourselves as gods. That's prideful. On the other hand, adapting, progressing, and showing humility leads to gratitude for all we have been blessed with.

Much of today's climate strife stems from trying to replace God. If you become indentured to a system of men, that system becomes hell, as history proves. Non-Christians will say I am being deceived or delusional, but my stance involves a *reduction* of my status. The other stance — a yearning to "save the planet" — is almost messianic. Think about the Pax Romana and the desire of Caesar to ensure world peace. He thought he knew better than everyone else and was even declared a god, which ultimately brought nothing but terror and hardship to much of the world. It's not too far-fetched to think the climate movement is a modern-day version of the Pax Romana, only instead of Caesar being worshipped, we are dealing with the worship of Gaia.

Interestingly enough, the Pax Romana coincided with a climate optimum, during which the weather conditions made it easier for Roman armies to conquer their foes. We are approaching a climate optimum now, but the portraying of it as a climate emergency is largely soft tyranny that's meant to scare the public — similar to how the armies of Rome scared their adversaries.

CO_2 was not driving the planetary climate then, and there is a great deal of uncertainty regarding what, if any, measurable effect it is having now. We have not yet reached the level of tyranny that the Romans enforced to ensure their Pax Romana, which consisted of putting to death or jailing anyone who dared speak against them. But we *are* seeing the shutting down of debate — and even calls for trials.

The Weaponization of Drought

CHAPTER 2

There are many spectacularly busted forecasts that climate doomsayers have gotten away with. The ones touting the dry summers of 2010-2012 as the beginning of a perma-drought were among the most egregious I've seen. Yet most of the media were mum when the exact opposite pattern took over in the ensuing years, though they did blame man-made climate change when the heavy *rains* showed up.

My direct opposition to these perma-drought predictions was stated loudly and clearly at the time. A couple of salient points emerged from their forecasts, the most pertinent being this: How did we have even worse droughts before with lower amounts of CO_2?

Climate-change doomsayers ignored the key reason for the 2010-2012 summer droughts. The corollary was cold water off the west coast of Mexico. Once that reversed, like clockwork it flipped the pattern over the U.S. The fact is, when the East Pacific cools, there's usually a bit of a downturn in global temperatures, as we saw in 2010-2012. *However,* the U.S. also turns dry. It's easy to understand why. When a large area of the eastern and tropical Pacific cools, the feedback leads to a lowering in global temperatures. Drier summers then prevail, as less water vapor reaches the country from the west. North of the tropics, it only takes small changes in source regions to produce big changes in the weather pattern.

The 2010-2012 southern Great Plains drought closely mirrored the 1952-1954 drought. Jim Dent, in his must-read book for any Texas Aggie fan, *The Junction Boys: How 10 Days in*

Hell with Bear Bryant Forged a Champion Team, has some vivid descriptions of how bad that three-year drought was. The drought culminated during the 1954 Aggie training camp in Junction, Texas. For me, the Aggies and the weather represent a common intersection of love. (Over the years, both have led to heartbreak. But what is love without pain?)

Let's look at these two droughts courtesy of NOAA's Physical Sciences Laboratory[1]. On the left is 1952-1954, and on the right is 2010-2012.

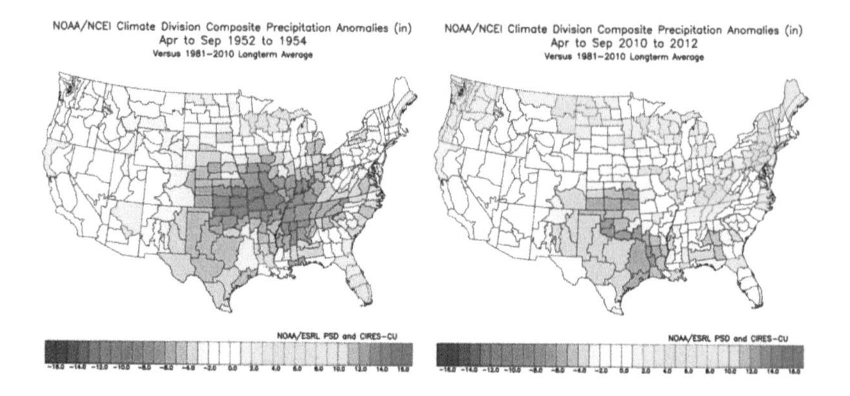

Would it surprise you to know that the 1950s drought was more widespread? The common denominator can't be CO_2, because it was much higher in the latter drought. Who knows — maybe more carbon dioxide resulted in the drought being *less* widespread.

Meanwhile, look at sea surface temperatures off the west coast of Mexico during the 1950s drought.

[1] Data/images provided by the NOAA/OAR/ESRL PSL, Boulder, Colorado, USA, from their Web site at http://psl.noaa.gov/

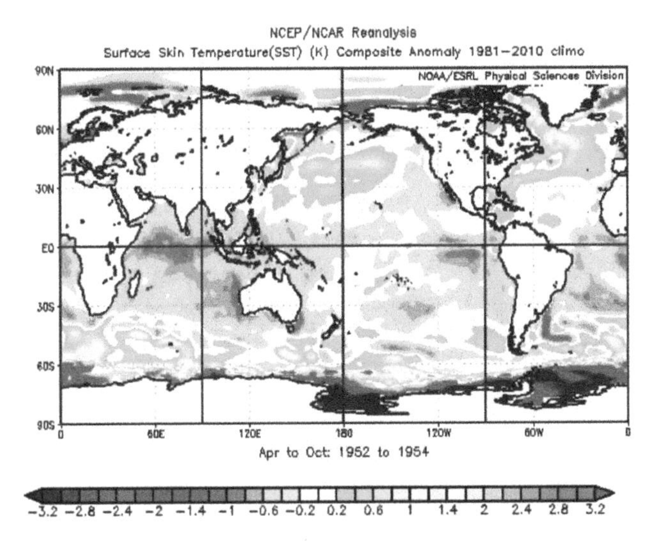

While the oceans overall were much cooler then — and that may have contributed to less water vapor and more widespread dryness over the U.S. — the same cool patch showed up in 2010-2012.

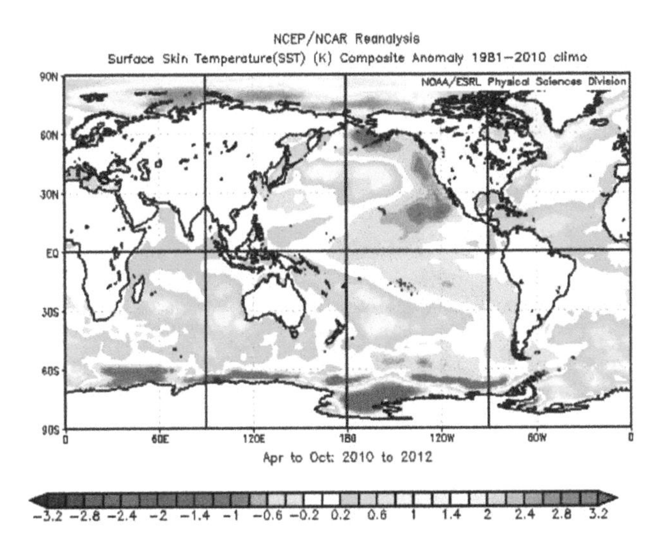

Once the temperature flip occurred in the oceans, guess what happened? Like clockwork, most of the U.S. turned wet, aided by extra input from the warm Gulf of Mexico and Atlantic. The graphics below show the change in sea surface temperatures from 2014 to 2019 (left) and the corresponding precipitation anomalies (right).

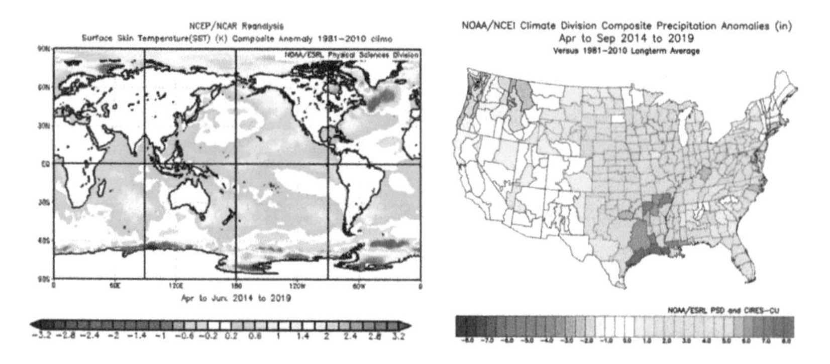

Clearly, those who were predicting a new dust bowl either didn't look at this evidence, or they did and simply decided to conceal it since it contradicted their agenda.

Of course, a few years later, the California pattern also flipped, and that too can be traced directly to a change in sea surface temperatures. This has nothing to do with CO_2, yet no one even questions it. Even in May 2014, this headline appeared in National Geographic: "Parched: A New Dust Bowl Forms in the Heartland."[2]

For those who think these weather patterns do hinge on CO_2, why instead of a dust bowl did it get so wet? You can't have it both ways.

[2] https://www.nationalgeographic.com/news/2014/5/140516-dust-bowl-drought-oklahoma-panhandle-food

The drought severity index in the 2010-2012 drought covered a smaller part of the nation relative to 1952-1954.

Here were 2010-2012 (left) and 1952-1954 (right):

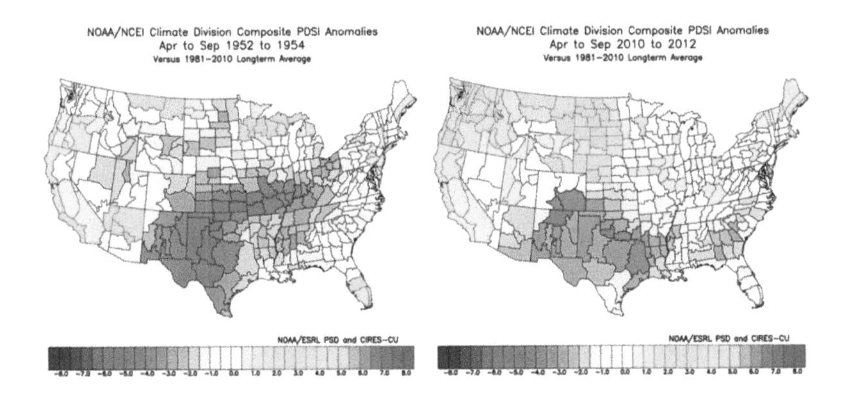

How would you feel about an *entire decade* of severe drought? Check out the 1930s.

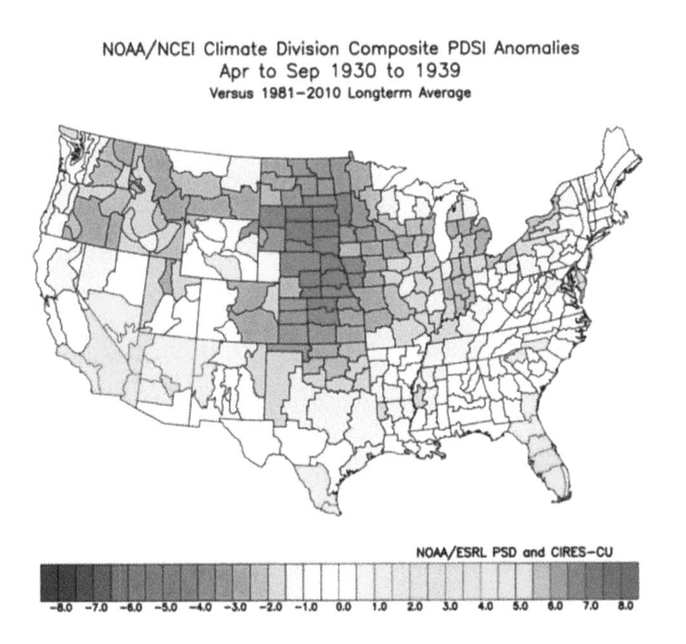

There's simply no other decade in U.S. history that can match the 1930s with the combination of heat, drought, and hurricanes (though the 1950s comes close). If we developed a scale rating major hurricane hits, drought severity, and heat, while the most recent decade may challenge the 1930s by way of warmth (due to elevated *nighttime* temperatures resulting from increased water vapor), the high-temperature extremes of the 1930s can't be matched.

Blinded by the narrow-minded focus on pushing an agenda, climate doomsayers couldn't care less about the cause if it doesn't fit their missive. One of the big keys to the kind of heat and drought they predicted to develop is the assumption that subtropical ridging (high pressure) would strengthen enough to warm the mid and upper levels of the atmosphere, resulting in less precipitation.

Here's what that means: When air is warm and moist, it's more buoyant. But condensation processes — from which we get precipitation — slow down if the upper levels of the atmosphere are warming. Currently, the oceans, due to natural cyclical processes, are in their warmer phases, so there is indeed more water vapor. However, the warming needed in the upper levels to create long-term drought isn't there. Again, blinded by the CO_2 light, doomsayers busted.

More warming means more precipitation, including snow, in the northern hemisphere, which gets returned to the oceans. A flip occurs when the oceans start to cool. All this takes a long time, but the point is that I'm seeing the embryonic stages of that process. This was taught in meteorology school in the 1970s, when the coming ice age was all the rage. That spectacular bust had people predicting the same kind of doom and gloom we hear now for global warming.

How even the most ardent believer in anthropogenic global warming doesn't stop to at least question these terrible forecasts shows me that the objective is to use the weather for an agenda. Hence the reason for the name of this book.

The Weaponization of Tornadoes

CHAPTER

3

The table below came from a NOAA paper entitled, "Normalized Damage from Major Tornadoes in the United States: 1890-1999."[3] Look at the top 10 most damaging tornadoes adjusted for inflation:

Rank	Date	Location	Raw	Adjusted
1	May 3, 1999	Oklahoma City, OK	1000	963
2	April 10, 1979	Wichita Falls, TX	400	884
3	May 6, 1975	Omaha, NE	250	745
4	May 11, 1970	Lubbock, TX	135	558
5	June 8, 1966	Topeka, KS	100	494
6	October 3, 1979	Windsor Locks, CT	200	442
7	May 27, 1896	St. Louis, MO - E. St. Louis, IL	12	380
8	April 3, 1974	Xenia, OH	100	325
9	March 31, 1973	Conyers, GA	89	321
10	June 9, 1953	Worcester, MA	52	311

Notice the dates. While this particular data is only good through 1999, it remains true that the bulk of the most damaging tornadoes occurred when the planet was *colder*. Why might a colder climate produce more extreme tornadoes? The Gulf of Mexico is always at the ready to supply warm, humid air. And if

[3] https://www.nssl.noaa.gov/users/brooks/public_html/damage/tdam1.html#Table2

it's colder across the mainland, the temperature gradient is more likely to produce extreme threats. On the other hand, if there's less cold air available, there tends to be less zonal potential energy. And if there's *too* much cold air, then you won't get the ideal temperature gradient either.

The Storm Prediction Center (SPC) is a classic example of taxpayer money well spent, as tornado deaths have *not* increased. This is a tribute to storm chasers as well as to the SPC, which coordinates with local offices to provide warnings to the public. In the NOAA table above, only one of the most damaging tornadoes occurred in the warm era — the 1999 Oklahoma City tornado. An outstanding summary from Axios[4] chronicling the top 10 deadliest tornadoes reveals the same thing. Only the 2011 Joplin tornado (ranked 7th deadliest) occurred in the warm era.

In 2012, there were only 939 tornadoes. Look what March, April, and May temperatures looked like (charts courtesy of NOAA's Physical Sciences Laboratory).

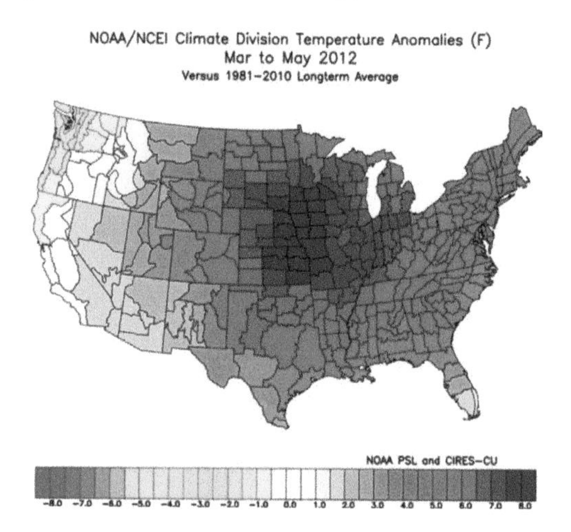

4 https://www.axios.com/extreme-weather-worst-tornado-outbreak-89dbf8f6-5c63-4593-8890-0cf944513350.html

2011 had 1,703 tornadoes. The cold was available for the clash. Here is the March-May temperature average:

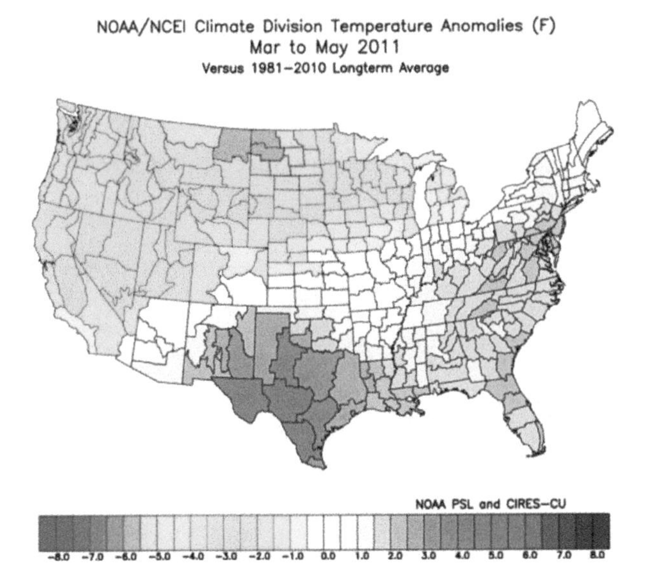

What's going on here? Today, almost every tornado not only has an eyewitness but is often caught on camera. There were not many cameras around to catch the tri-state tornado, for example. From the aforementioned Axios article:

> *The single deadliest tornado to ever hit the United States, the "Tri-State Tornado," killed 695 people and injured 2,027 others in Southern Missouri, Illinois, and Indiana in 1925. The tornado went on for 219 miles, making it the longest ever recorded.*

Today, there is video available almost all the time — if not from storm chasers, then from the general public. Radar can also identify weaker, more rural tornadoes, similar to how satellites catch tropical systems in the middle of nowhere. Remote tropical storms and hurricanes might get named and added to the list today, but many of those same storms would have been undetectable before the satellite era. This is one way that climate-change extremists are weaponizing both tornadoes and hurricanes.

The following chart, derived from NOAA's National Centers for Environmental Information[5], reveals a *decrease* in strong tornadoes.

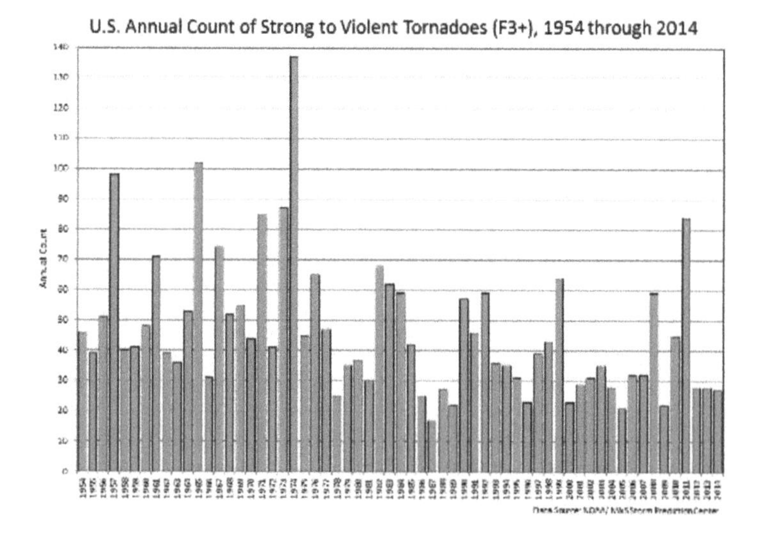

But that little fact doesn't fit the prevailing missive. Yet the extremists demonize those who point out their hypocrisy and march in lockstep with a complicit media.

Tornadoes are dramatic events that can destroy lives in a heartbeat. But to claim they are getting worse because of man-made climate change is flat-out wrong.

---

[5] https://www.ncdc.noaa.gov/climate-information/extreme-events/us-tornado-climatology/trends

CHAPTER 4

The Weaponization of Floods

Below you will find the top 10 flood disasters in U.S. history by way of death, courtesy of Joseph Kiprop[6].

1. Johnston, Pennsylvania, 1889: 2,209 deaths.
2. St. Francis Dam Failure, Los Angeles County, 1928: 431 deaths.
3. Ohio River Flood, 1937: 385 deaths.
4. Dayton, Ohio, 1913: 360 deaths.
5. Great Mississippi River Flood, 1927: 246 deaths.
6. Black Hills Flood, South Dakota, 1972: 238 deaths.
7. Los Angeles Flood, 1938: 115 deaths.
8. Columbus, Ohio, 1913: 90 deaths.
9. Laurel Run Dam Failure, Pennsylvania, 1977: 80 deaths.
10. Austin, Texas, Dam Failure, 1911: 78 deaths.

In July 2019, USA Today[7] listed the top 30 worst floods that include impacts from hurricanes. Both USA Today's list and the one above rank floods based on death counts.

Hurricane Harvey, while 30th in deaths, was the costliest, while Hurricane Katrina resulted in a much higher death toll but only ranked 2nd in total dollars. Both struck populated areas.

---

[6] Kiprop, Joseph. "The Worst Floods in US History." WorldAtlas, Oct. 13, 2017, worldatlas.com/articles/the-worst-floods-in-us-history.html

[7] https://www.usatoday.com/story/news/weather/2019/07/17/worst-floods-in-american-history/39692839/

More buildings and concrete were contributing factors, as both disrupt natural drainage.

When filtering out hurricanes, the most recent deadliest flood occurred in 1977 in Pennsylvania. Before that, the most recent was in 1972, also in Pennsylvania. Before those two, the deadliest floods were pre-1950s. Some of them were the result of dam failures. Overall, though, the most extreme events haven't reoccurred. In fact, so much time has elapsed between exceptionally deadly flooding events that we should expect something soon. The Carolina flooding from Hurricane Florence was devastating, and I'm worried about the eastern and southern states this year given yet another wet spring and the prospect of a major hurricane season.

We can conclude a few things.

- If today's weather is "worse," we've done a great job adjusting to it.
- If more damage is occurring today, that's because so much more real estate is in the way.

But is it raining more as a whole? In some places, yes. The question is, why? Oceans are the biggest outsourcing mechanism for water vapor and CO_2. Look at the change in sea surface temperatures over the last 35 years (graphic via Weatherbell.com):

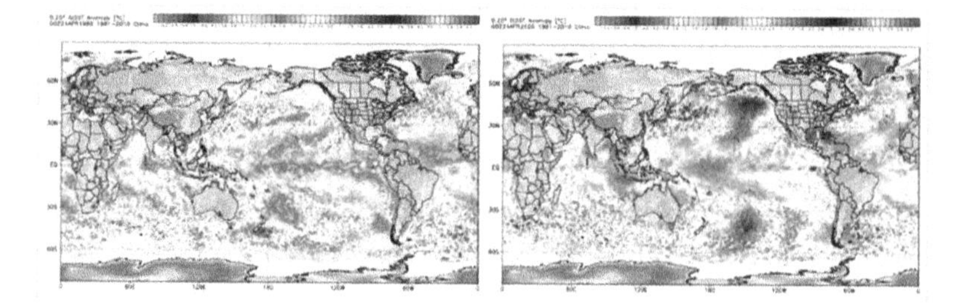

The added ocean warmth has put more water vapor in the air. If the atmosphere was heating up enough to counter this influx, then the opposing side would indeed be right about widespread areas of drought. In fact, the next time the tropical Pacific cools, like it did in 2010-2012, we may see more droughts. If the western Atlantic and Gulf of Mexico cool, that too will have a big impact. But the point is that, right now, because the mid and upper levels of the atmosphere are not sufficiently warming, there's enough instability to produce more rain.

Warm oceans warm the air, so it's probably not a coincidence that global surface temperatures have risen in tandem with the oceans. But is surface warming also linked to CO_2? I would argue it's not.

Even though the left is piggybacking on climate change to weaponize COVID-19, it seems blissfully unaware that the shutdown is having *no discernible effect* on the seasonal rise of CO_2, while other so-called pollutants have slowed or dropped.

The graphic below is from the University of California San Diego's Scripps Institution of Oceanography[8].

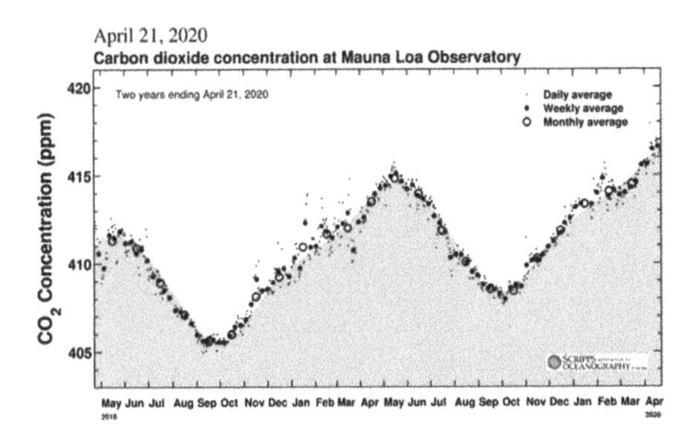

[8] https://scripps.ucsd.edu/programs/keelingcurve/wp-content/plugins/sio-bluemoon/graphs/mlo_two_years.png

The COVID-19 shutdown was not accompanied by any significant slowdown in CO_2. There was a flattening for a time, but we saw the same thing last year when the economies of the world were running full blast. Clearly, arguing for economic shutdowns to stop CO_2 from increasing doesn't hold water. You can't have it both ways. You can't shut down commerce and then claim you were right about carbon dioxide when the facts reveal the opposite.

Why is there a seasonal downturn? Because the northern hemisphere greens up, and plants take CO_2 out of the air. As you can see in the day-to-day variance, the measurement is sensitive. If man's input was having an effect, it should show up.

Where's all this CO_2 coming from if not man? Recall that the oceans are the biggest source of consistent input of carbon dioxide. At the very least, this suggests that until the oceans reach an equilibrium, they will continue to provide a positive input of CO_2. You can tie temperature surges to events like Super El Niños, but you can't tie them to CO_2.

There's an idea out there that CO_2's rise is contributing to warming above the surface, which then limits cooling. The aggregate effect is up for debate. I believe it has *some* impact, but far less than what doomsayers claim. But let's just assume they have a point. So what? CO_2 continues to rise despite the coronavirus shutdown, meaning it's not (completely) man's fault. Therefore, let's do what we always do: Adapt to what's coming.

If the oceans are causing the rise in CO_2, that means the real driver is natural. Which makes sense given what we are seeing with the increase in the most important greenhouse gas — water vapor — which circles back to precipitation, and precipitation to floods.

The greater storm damage we see today is because our progress has landed us in harm's way. So, let's adapt. One way to adapt is by being prosperous and fruitful to better prevent problems, or at least slow them down. If there's a silver lining to COVID-19, maybe it's that the associated shutdowns will open some eyes. In the meantime, prepare for rampant hysteria about any and all flooding events. Just have an open mind about what's causing them.

CHAPTER

The Weaponization of Hurricanes

If you give it a name, you increase its fame. The drivers of the climate-change agenda know this, and so does the media. I have nothing against naming storms. In fact, the Travellers Weather Service was naming winter storms in the 1970s. I named them in the '70s after radio personalities on the stations I was on as an undergrad. It was fun.

But the increase in the number of named storms lately has been used as a form of weaponization by people who are suggesting hurricanes are getting worse. The fact is, we are *seeing* more storms today because of an increase in detection. A classic example is comparing the 1933 and 2005 hurricane seasons. (Maps courtesy of the National Hurricane Center.)

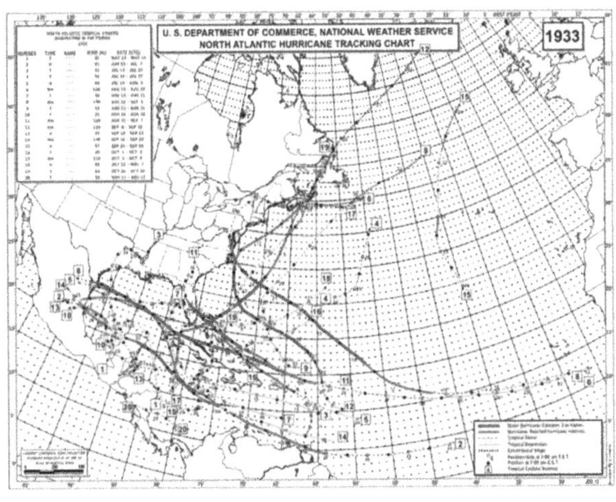

The 1933 Atlantic basin hurricane season, revised.

The revised map above is different from the initial map, which you can see below.

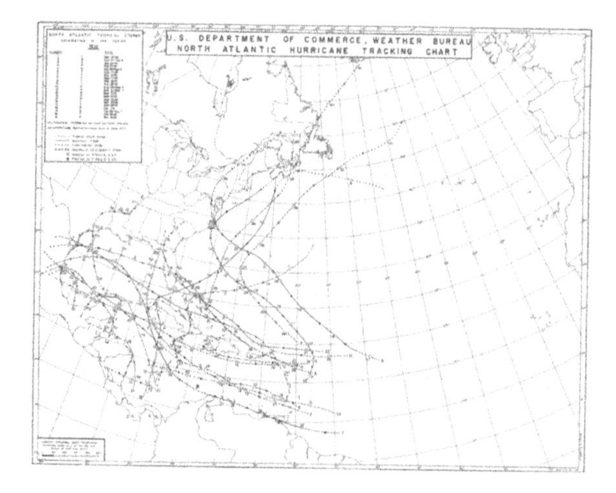

Notice how in the "revised" analysis, storms were "discovered" farther east and even in the middle of nowhere. This is rather interesting since, until reanalysis, the information available then said there was nothing there. In any case, the 1933 hurricane season was just as active as 2005. The difference is that in 2005, many storms were seen in areas we could not have known about in 1933.

Here's 2005:

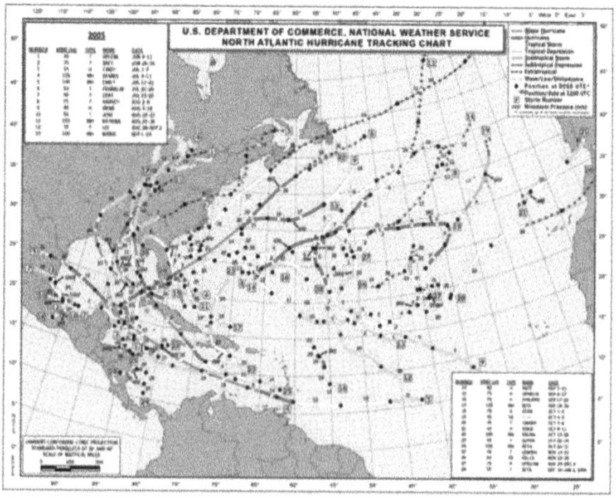

The way to map these storms objectively would be to have two maps — one for storms that gain tropical cyclone status over water warmer than 26°C, which is the long-accepted threshold, and the other for storms that we would have never seen before and, quite frankly, are of no concern to land. (Though for shipping interests or perhaps Europe, that's a different story.)

We know the number of storms is inflated. But when we look at the number of major hurricane hits on the U.S. coast, *it's actually going down.* The Accumulated Cyclone Energy (ACE) index, which is a method to objectively quantify the strength of the hurricane season, reached its per-storm peak of 10 in the 1950s. In the most recent decade, it's fallen to 8.3. Moreover, major landfalls have dropped. From 1931-1960, 29 storms of Category 3 or greater hit the U.S. Hurricane Donna (1960) made three different landfalls and unleashed hurricane-force winds on each state from Florida to Maine, something that no other storm has come close to accomplishing. Since 1990 — and remember, all the global-warming hysteria really got started in 1988 with the congressional testimony of James Hansen — we have only had 16 major hurricane hits. Of these, only four made landfall at full strength, and those were the ones developing relatively close to land.

As wild as the 2004 and 2005 seasons were, many of the storms were actually off their peak at landfall — a trait we have been seeing quite a bit of late. (I will explain why I think this is a bit later.) From 1915-1919, the Gulf was raked by seven major hurricanes in five years, all of them near max intensity at landfall. Eight major hurricanes roared up the East Coast from 1954-1960, including three in 1954. Two of them, Carol and Edna, occurred within 11 days of each other — in New England, of all places! The storm surge from the 1938 hurricane flattened The Hamptons, and the storm surge into Narragansett Bay put Providence, Rhode Island — which sits 12 feet above sea level — under 13 feet of water. There were reports of wave heights of 30 to 40 feet, which when combined with the storm surge added up to an almost incomprehensible amount of water. What's more impressive in terms of extremity: the 1938 hurricane, or Katrina, which traversed an ingredient-friendly Gulf of Mexico and flooded New Orleans, much of which sits *below* sea level?

By the way, why people who support the climate-change agenda would build mansions in places that have been devastated by hurricanes is beyond me. Case in point: Former President Obama's mansion on Martha's Vineyard sits three feet above sea level at the north end of a funnel-shaped bay that has only a five-foot barrier island to stop storm surge. He can buy his house wherever he wants. I applaud his success. Just don't blame climate change when hurricane monsters from the past return.

Hurricane Carol in 1954 sent 12 feet of water into Rhode Island. It got so bad that the senator from Rhode Island, Theodore Greene, was instrumental in getting a "radar fence" built on the East Coast so that New England would be prepared.

Hurricanes are being used as a weapon to dupe an unsuspecting public that has not been shown how extreme the weather can be. As a result, they sit in awe and swallow the idea that today's weather events are unprecedented while clamoring for the supposed safety being offered by extremists whose "solutions" are simply a smoke screen for a political agenda.

We are acquainted with Hurricanes Harvey and Michael — storms that hit while ramping up, which was forecasted by my company several days in advance. But lost in the ruckus is the fact that Hurricane Irma was well off her peak when she hit Florida.

Florence was inspiring talk of a "Category 6," but she eventually weakened. Dorian smashed the northern Bahamas, but weakened coming to the U.S. Matthew, Irene, Katrina, Rita, Wilma, Ike — I can go on and on. All became *weaker*, not stronger like we saw before the era of global-warming panic.

There's a simple reason for why damage totals are increasing despite storm intensities not being as bad: More people are living in harm's way. A classic example is a storm like Hazel, which hit Myrtle Beach as a Category 4 in October 1954. Myrtle Beach was not nearly as built up then as it is now. If Hazel comes again, the damage will be exponentially worse. Yet some people are saying the weather is worse than ever. Somehow, they believe that we can construct a man-made utopia that eliminates risk. Here's a fact: If you live on a beach, you can get hit by a hurricane. Live there long enough, and *you are going to get hit* by a hurricane. There's also a good chance it will destroy your property. Sixty years ago, that likely wouldn't be the case, because your real estate likely wasn't there.

Do we clear the beaches? Do we say, "Don't build"? No. We accept the risk and responsibility, we adapt, and we don't point fingers at anything but nature. I don't mean to sound harsh, but the late Dr. William Gray said he believed a quarter-trillion-dollar storm was likely to hit the U.S. That was back in the 1970s, when a coming ice age was all the rage. But Dr. Gray was saying the Atlantic Ocean would go into a warmer phase, and we would pay a price. He was spot on then, and he is spot on now.

Side Note 1: Miami Beach is built on a man-made sandbar. Cyclical rises and warming of our oceans have always occurred along with cooling. The idea is not to abandon Miami Beach but to understand the naturally changing climate and adapt to it.

Side Note 2: Globally, there has been no increase in the ACE index. With increased satellite coverage, once hurricane seasons get underway, there's always something happening somewhere.

Side Note 3: The top 10 deadliest global tropical disasters[9] occurred in 1970, 1737, 1881, 1839, 1584, 1876, 1897, 1975, 1991, and 2008. The vast majority of those were in the pre-warming era.

Ironically enough, hurricanes today may be weakening as they approach land because of global warming. But it's *disproportionate warming*. There's been more warming in the northern latitudes than over the tropics. Therefore, while more storms may develop farther north in the oceans, they are likely to stay out at sea — just names and numbers. That's because the sea-level pressure patterns during the hurricane season have changed. Hurricanes need high pressure around them to increase convergence and inflow into the center (or "eye"). The processes that take place with relatively small pressure changes over warm water are much more significant than over cooler areas. A mature cyclone, which will naturally tend to expand in size with age and weaken at its core, will do so even more if pressures are lower to its north and the water is warmer than average. It will pull the convergence away from the center, weakening the core.

I'm sympathetic to some scientists' belief that global warming is resulting in slower storm movement and enhanced rainfall. But I also feel we need more research. Hurricane Harvey's stalling was weaponized as a sign of global warming, when nothing was further from the truth. The argument has been that stronger

[9] Liu, Defu & Wang, Fengqing. (2019). Typhoon/Hurricane/Tropical Cyclone Disasters: Prediction, Prevention and Mitigation. Journal of Geoscience and Environment Protection. 07. 26-36. 10.4236/gep.2019.75003.

ridges slow these storms, but Harvey got trapped by a powerful cold trough. It was a classic case to make my point about high-impact storms. Cooler-than-average air and higher-than-normal surface pressure ahead of the storm meant convergence and inflow over the Gulf of Mexico as the storm developed was increasing — until, of course, it made landfall and the trough captured it, at which point the immense amount of moisture rained itself out in a band that stalled around Houston. The condensation processes were *enhanced* by the cooling with the trough.

Hurricane Florence was a better argument for the subtropical-ridge theory. However, it weakened so much despite the ridge that, relative to the strength it was, the storm was a has-been by the time it reached land. The hysteria that was being whipped up at its peak included talk of a "Category 6."

For the record, I think there *is* a way man can affect the intensity of hurricanes. I think we should restart Project Stormfury by seeding strong hurricanes in the 36 hours leading up to landfall. The theory is simple: The stronger the storm, the more ideal it needs the environment to be to remain strong. Disrupt it with massive round-the-clock seeding and you may knock it down a category. The last storm to be seeded, Debbie in 1969, lost 31% of its peak wind speed the first day and 18% the second day. If I'm right, Project Stormfury would be a relatively inexpensive but powerful hurricane remedy.

Note: A reanalysis in the 1980s called into question the results mentioned above. But the bottom line is:

What do we have to lose? Nothing. Yet I suspect the weather-is-worse-than-ever crowd won't go for it.

A chapter like this could go on forever. The main point is that hurricanes have become another display of nature that's being dragged through the mud by people who have an ends-justify-the-means mentality. All of the storms I mentioned could naturally have been a bit stronger. We need to accept the risks — and adapt. Hurricanes should never be a political weapon. Yet that's what they have become.

The Weaponization of the Global Temperature

CHAPTER 6

The oceans have warmed dramatically. Via Weatherbell.com, here's the difference between the 1980s and now:

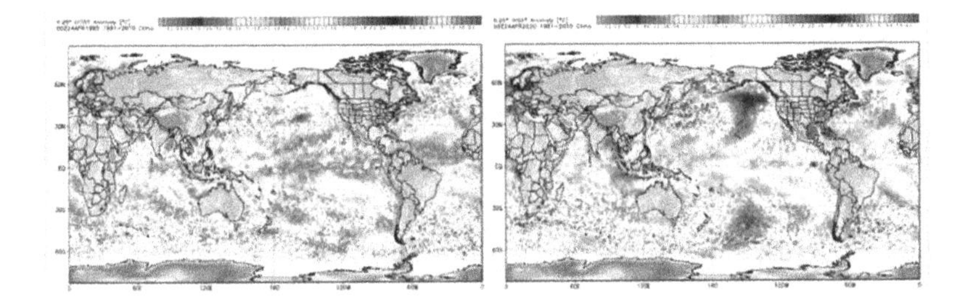

The oceans, being the largest reservoir of CO_2, must be outsourcing it into the air. Which makes them a vital player in the global rise in carbon dioxide. My hypothesis is that because the oceans are so warm, they are constantly trying to find an air-ocean equilibrium.

What is causing ocean warming? Several factors may be at play.

For starters, increased sunshine over the deep tropics may be influencing sea surface temperatures. (The temperature and water vapor charts below are courtesy of NOAA's Physical Sciences Laboratory.)

Here is the composite anomaly for humidity at 400 mb (roughly 23,500 feet) from 2006-2017:

Another clue may be underwater warm thermals. The state of the oceans today is a product of decades (or more) of action and reaction. That doesn't eliminate CO_2 as a factor, but it does suggest there are other facets.

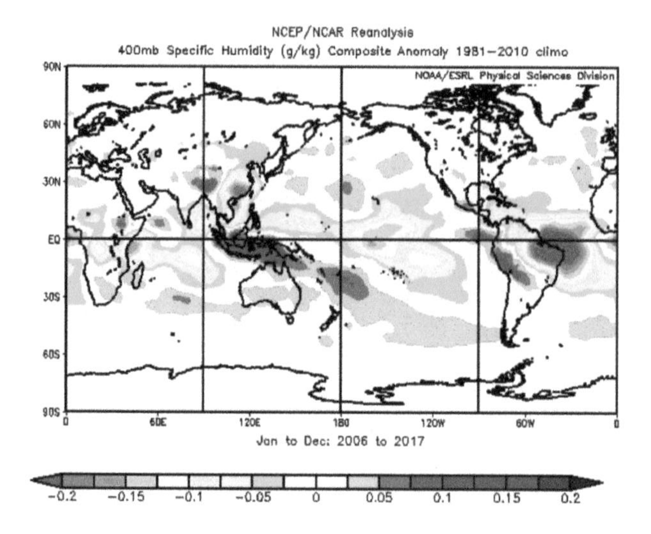

Water vapor being the number one greenhouse gas is the most important factor, and the warmer oceans are outsourcing it. More warming is occurring in places that are cold and dry during the winter season. This is seen clearly in the temperature anomalies. Here's the December-February change from 2006 to 2017:

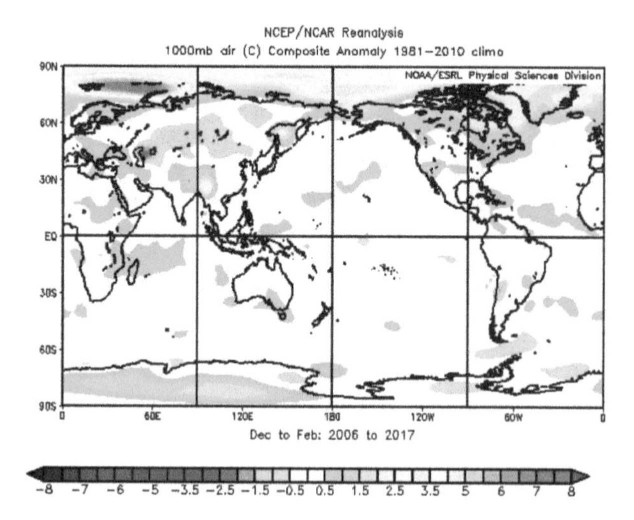

Temperatures in the Arctic are skewed heavily by winter seasons. The same trend is evident in the Antarctic. The following map is courtesy of the Danish Meteorological Institute.

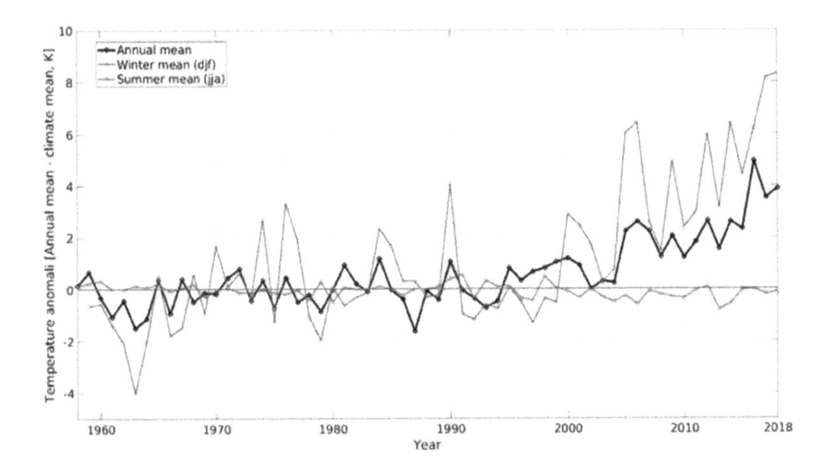

Notice the red line, which represents summer temperatures. Why no increase? Two things: The process of ice melt takes heat out of the atmosphere, and the increase in water vapor is not enough to affect temperatures, unlike during the winter.

Speaking of water vapor, I want you to look at the relationship it has with temperatures. In the chart below, notice how much more water vapor is needed as dew point temperatures (represented in increments of 10 degrees) get warmer. Slight increases in water vapor where it's cold and dry result in much bigger dew point temperature rises.

As you can see, an increase of .09 grams/kg of water vapor at −40°F correlates to a 10-degree higher dew point temperature. At 70°F, a 10-degree dew point temperature rise needs an increase of 6.48 grams/kg of water vapor. The higher the dew point, the higher the minimum temperatures in general. We're seeing warmer nighttime lows across the U.S., which is linked to water vapor, not CO_2.

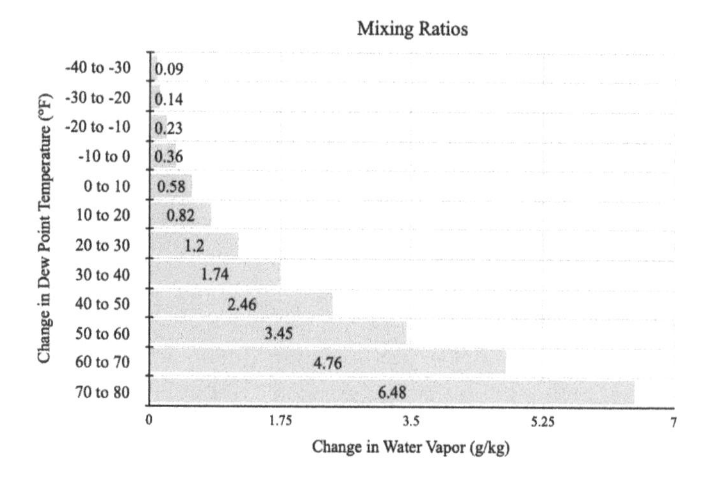

Let's look again at winter temperature anomalies from 2006 to 2017 (left) and compare them to the increase in water vapor anomalies (right) at the same time.

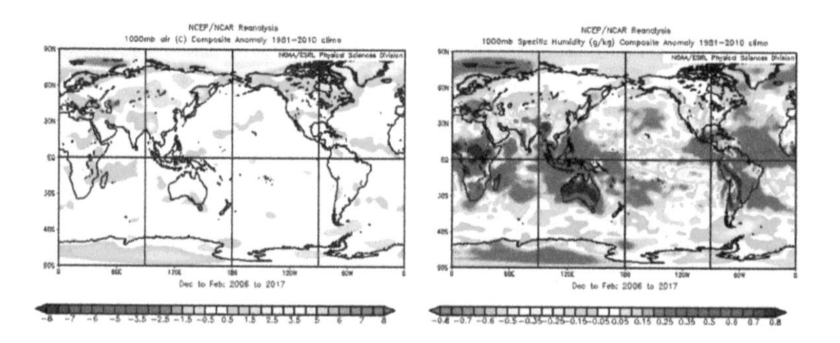

Despite massive water vapor increases in the tropics, the warming is more pronounced in the Arctic. Moreover, in a large part of the planet where life thrives, the temperature rise has been much less noticeable. *The warming is distorted.*

Even where it's warmed the most, the increase is dull enough and the rate is slow enough that we can adapt to it.

I have a theory about the rise in global temperatures. A closer examination reveals that Super El Niños are correlated with temperature step-ups. Remember the mixing ratio chart above. More water vapor is needed to keep dew point temperatures moving upward. So the amount of water vapor released in these Super El Niño events must be immense if water that is normally 83°F is now 87°F. While this water vapor is quickly dispersed, the process leads to a net rise in the amount of water vapor in the air. And because the oceans are warm, these events supply new temperature plateaus, which I'll demonstrate below.

The following graph[10] of satellite data — the gold standard for temperature measurements — is from the University of Alabama in Huntsville's Dr. Roy Spencer and Dr. John Christy. I have overlaid it with the aforementioned temperature plateaus. You'll notice very clearly the step-up effect of Super El Niños.

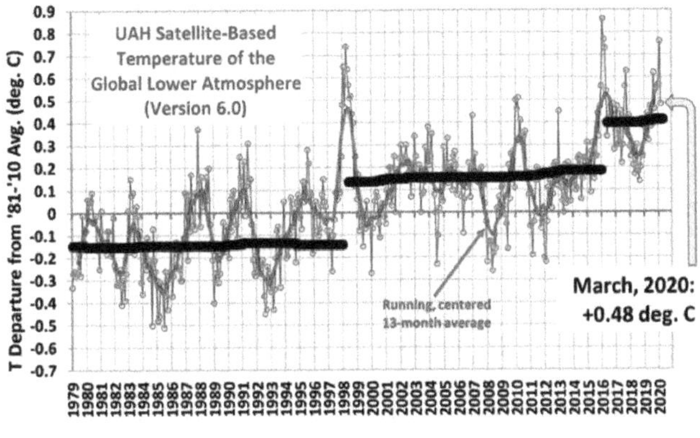

(Note: The black lines were added by Joe Bastardi.)

[10] https://www.drroyspencer.com/2020/04/uah-global-temperature-update-for-march-2020-0-48-deg-c/

The ups and downs correlate nicely not with the rise in CO_2 but with El Niños and La Niñas. More evidence that the oceans are the big driver.

Am I right? Well, that's up for debate. But in a way, that's the point. There *should* be debate. At the very least, this evidence offers some interesting implications.

I will conclude this chapter by highlighting a classic agenda-driven method. I am not into personally attacking people. It does me no good. But their ideas? Well, those are fair game.

On April 22, scientific historian Naomi Oreskes tweeted "Meanwhile…" while linking to a Washington Post article entitled, "Relentless record heat roasts south Florida while most of the Gulf Coast also is cooking." Said article featured this map from Weatherbell.com:

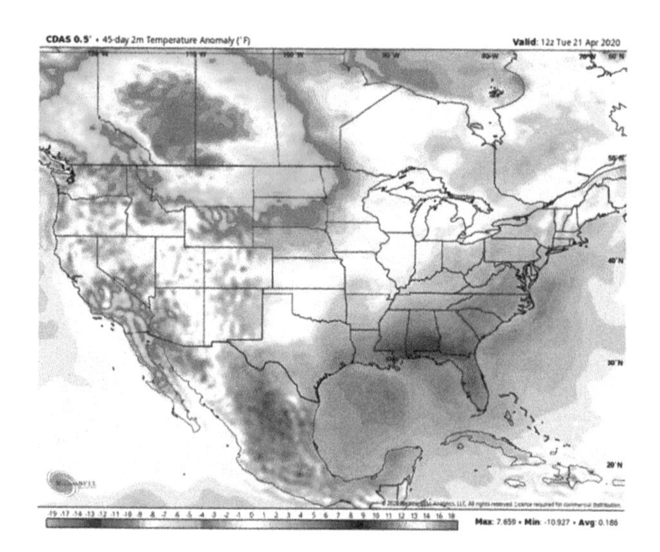

Ms. Oreskes is seemingly weaponizing a one-day snapshot and in no way addresses the overall picture.

It's certainly impressive that Miami hit an all-time April record high on the 20th. The warmth in general over the Southeast was a function of very warm water in the Gulf and western Atlantic. (The following charts are also from Weatherbell.com.)

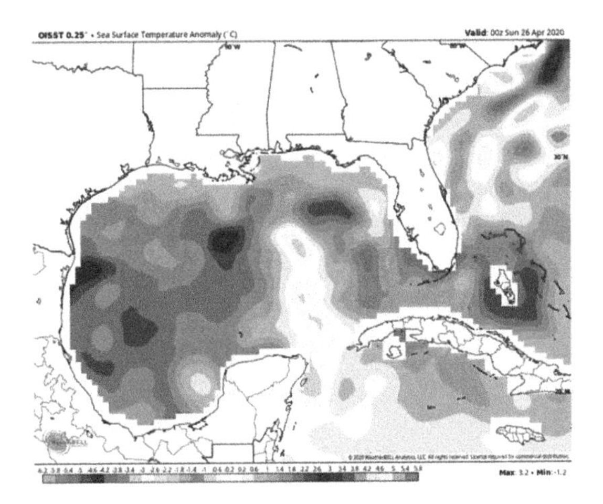

But a one-day snapshot doesn't do justice to what was happening nationally on a much larger scale.

Here were April 1-24 temperatures:

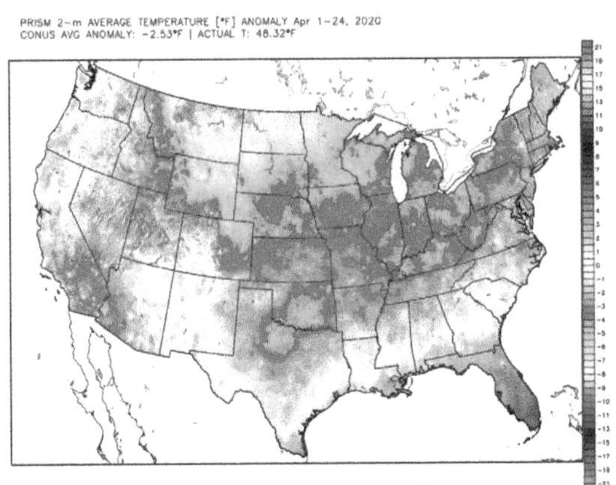

About 5% of the country had above-normal temperatures. Record lows outstripped record highs in April.

This is a typical tactic. In "Dancing in the Dark," Bruce Springsteen said: "There's something happening somewhere." Climate doomsayers know that on any given day, the weather somewhere will do something wild. That's the nature of the system. A willing media parrots what it's told and refuses to show the big picture, just like how the media refuses to show that much of the warming is occurring in the coldest, driest areas of the world.

It's a form of weaponization, and the supply line is the media, which refuses to look at anything that counters this missive.

CHAPTER 7

The Weaponization of "Global Weirding"

It's becoming increasingly popular among climate-change extremists to say that extreme cold spells occur because humans are making the climate weirder.

They say beauty is in the eye of the beholder. Well, so is weirdness. Those with a climate agenda gravitate toward an explanation in which everything that happens is a "sign" of what you predicted would happen.

Many years ago, I told Fox News anchor Neil Cavuto that if it snowed cheese in New York City, climate extremists would claim it's because of global warming. (Hopefully it would be a keto-diet-friendly mix of provolone and parmesan.) As absurd as that statement was, it made the point that even back then they were up to no good.

At the turn of the century, the idea that snow would become a thing of the past was making the rounds. But from 2003-2006, New York City experienced four consecutive winters of 40 inches or more of snow. Considering the average is around 25 inches, this was unheard of. So, naturally, climate extremists had to deflect.

The "bomb cyclone" and "polar vortex" hype really rankled me. Why? Because those terms were being used *in the 1970s.* In fact, being the weather nerd that I am, I wrote a song called "Bombogenesis in the Southeastern U.S." to the tune of "I Am A Rock." Snow nuts like myself know that rapidly developing storms — called "bombs" back in the day — dump lots of snow. Evidently, the media has no idea how long the term "bomb cyclone" has been around. So-called journalists run right into the hype and then say, "Man is wrecking the climate."

The "polar vortex" came a-calling in the late 1970s, during which time the planet was colder than it is today. But there was still warming over the Arctic that led to the same phenomenon that we saw more recently. Only now it's being attributed to man-made global warming.

As someone who relies on the past to frame the future, no stretch of U.S. weather was as extreme as the 1930s to 1960s. Honestly, everything from hurricanes to droughts to extreme heat to extreme cold during that period make the past several decades seem rather boring. But you'd only be able to acknowledge this *if you know your weather history.* For some people to pretend as if today's events are a sign that man is responsible for "weird" weather certainly has me questioning their methods.

Let's examine the last few times the so-called polar vortex paid a visit to the U.S. A single-day snapshot is limited and therefore inadequate. Averaging out the pattern over 10 or more days provides a much better depiction. I like using 20-day increments for 500 mb heights, so that's what I'll use below.

Here are the anomalies for the 2014 (left) and 2019 (right) Arctic outbreaks (all maps courtesy of NOAA's Physical Sciences Laboratory):

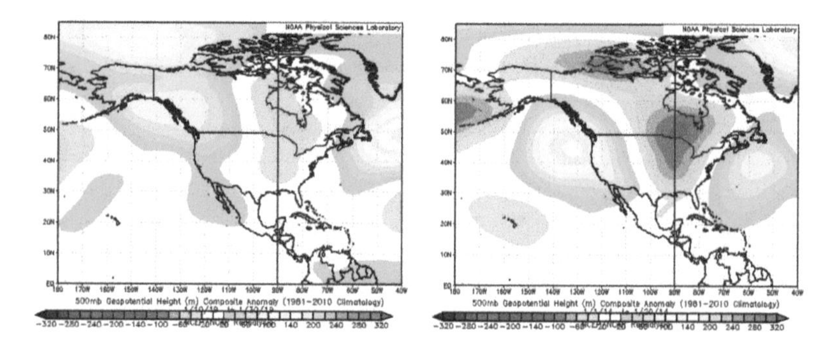

Now look at 1977. The blocking along with the intensity and length of cold over the U.S. was much "weirder."

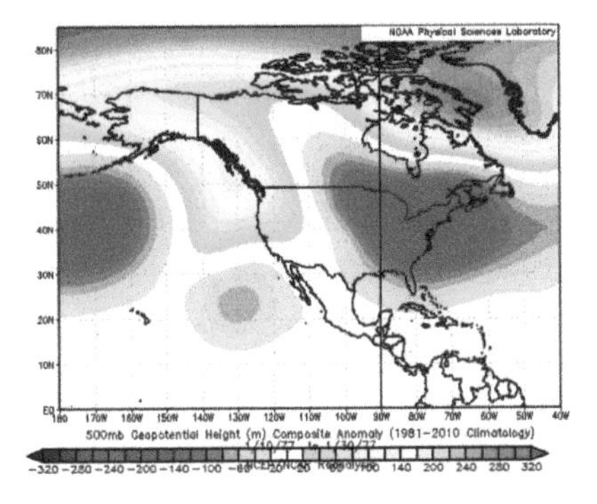

The Arctic was also "on fire" — another term for above-average temperatures — in 1977, even though it was still frigid there.

We predicted a late spring this year, and I think we can all agree that for the nation east of the Rockies, the lack of winter was followed by a late spring. Is that at all weird? Not to me, because my company was forecasting this very thing based on pattern recognition.

How extreme was the cold? If it was the most extreme case in history, that would be one thing. But we've actually experienced an even colder early spring outbreak— before the age of "global weirding." Below was the temperature anomaly for May 1-15, 2020 (left) and for the same period in 1966 (right):

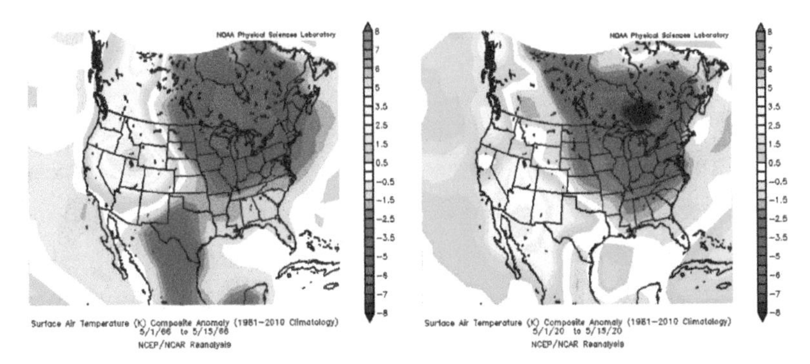

Which is more impressive in terms of extremes? The fact is that during an era of lower CO_2 levels, weather occurred that was just as "weird" or weirder than what we just saw. By the way, back in the 1960s and '70s, people claimed an ice age was coming. Some people on my side of the issue are claiming this now, and I don't like it. I point out to them that it's just nature being nature.

Until the oceans cool, we are not going to cool down globally. Will that lead to weirder weather? Well, if you don't know your weather history, you may be inclined to think so.

I don't know a single person in the weather business who loves extreme weather more than me. Quite frankly, this love of extremes occasionally fills me with guilt. But I want you to think about something. No one cares if the weather is tranquil. Why would I want the weather to be boring when people pay me to predict the big events? I would love to say, "Yes, this is weirder and more extreme than ever. Come to me for a forecast." But as guilty as I feel just loving extreme weather, that would feel even worse … because I don't believe it.

I reference my father quite a bit. He taught me that the foundation you stand on today to reach for tomorrow was built yesterday. You must ensure that the foundation is bedrock, not quicksand. The strength of the foundation depends on how much work you put into building it. In the case of climate change, that means researching the past.

I used to hate it when my dad disagreed with me on any given topic, because he inevitably turned out right. He knew more because he had more experience. I remember him telling me countless times how bad the weather was in the 1930s, '40s, and '50s. He would relay stories about my aunt being born in a snowstorm (my grandfather had to carry the midwife through the streets of Providence between mountainous drifts of snow) and the 1938 hurricane that put Providence under water (horns from submerged cars would blare out an otherworldly noise while the National Guard tried to stop people from looting jewelry). There were many, many stories. I used to think, "Sure, whatever." But as I got older, I studied all that he told me. It was wild. And guess what? If you dig in, you would also see that it's wild.

I think part of society's problem today is that we tend to elevate ourselves based on personal experience rather than by analyzing history.

Is the weather weirder? Are we seeing signs that man is wrecking the climate? Or is this rhetoric an agenda-driven missive that seeks to exploit the natural fear that comes from ignoring the total picture? I believe it's nature being nature. Perhaps we should acknowledge this, treat nature with awe, and accept the challenges she poses.

The Weaponization of Models

CHAPTER 8

The following graphs are from Dr. John Christy, University of Alabama in Huntsville.

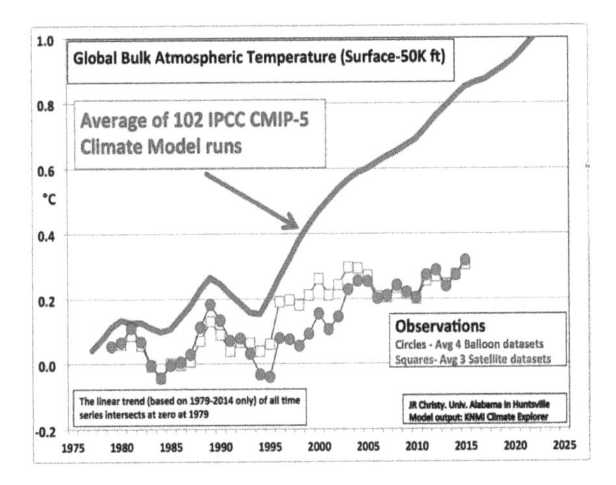

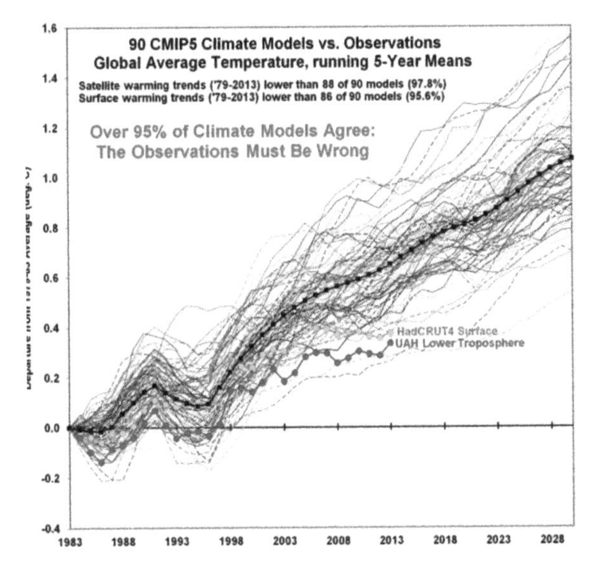

These graphs depict all the model temperature projections along with observed satellite temperatures. There *has* been a jump in temperatures over the last several years, which isn't depicted on these graphs, but that jump is likely linked to the 2015-2016 Super El Niño.

Many scientists, including Dr. Pat Michaels and Dr. Christy, are demonized for pointing out that observations are lower than what the models suggested. What's more, some of their antagonists claim the models are right. To some extent, they are, but not to the extent of a climate emergency. Any variation from the accepted missive means you get hammered.

This is very odd to me. Dr. Michaels and Dr. Christy, along with many other climate skeptics, have advanced degrees directly related to the climate issue. How is it that their educational achievements and scientific knowledge don't carry the same weight? Perhaps because this isn't about science at all. It goes without saying that people who have proven themselves by attaining prestigious degrees deserve equal treatment. How can a skeptic with a PhD be so wrong, yet people with a similar pedigree who toe the line are given a free pass?

Let me explain my reasons for model skepticism. I don't have a PhD in meteorology. What I have is 45 years of experience averaging well over 3,000 hours of work a year. Every day for the last 10 years, I have produced a product for use on a global scale. I did the same thing in the preceding 10 years at my former job, not because I was forced to but because I love it. If I didn't have to sleep, I wouldn't. I would instead stay up watching the weather. The tremendous advancement we have made today in the field of meteorology simply stokes the fire. I am a blue-collar laborer of Italian heritage. I have no illusions about that. But what I have seen and observed leads me to believe that models, though better today, still aren't perfect and can make large errors. If policy is based on modeling, it can lead to a cure worse than the disease. Gee, where have we heard that of late?

When feeding information to a computer model, the initial conditions are very important. Unfortunately, some analysts are cooling the old data, which creates a greater temperature disparity over time. But their information is based on reanalysis, whereas the satellite era is based much more on observations. Pre-satellite-era data for large areas is either estimated or not there at all. That's why I use satellite-era temperature observations. Granted, satellite-derived temperatures certainly are warming, but this can be linked to the warming oceans.

There are other factors as well. The huge argument is between runaway warming and the natural self-limiting process. This is described in Le Chatelier's principle, something us old geezers learned about when the ice age was all the rage in the 1970s. Lexico.com defines Le Chatelier's principle as: "A principle stating that if a constraint (such as a change in pressure, temperature, or concentration of a reactant) is applied to a system in equilibrium, the equilibrium will shift so as to tend to counteract the effect of the constraint."

In other words, there is a limit to the amount of warming that can occur. This principle essentially says that, because of the nature of heat, it's harder for temperatures to rise the further away that warming gets from a natural set point.

Notice the word "counteract." What does that mean? According to Lexico.com, it means: "Act against (something) in order to reduce its force or neutralize it." The system does that naturally.

My dad, who went to Texas A&M and graduated with a degree in meteorology at age 35 in 1965 (he went after going into the Army, getting married, and working), told me repeatedly that weather and climate are nature's eternal search for the balance she cannot have simply because of the design of the system. It's in constant conflict — a classic example of chaos theory. A February 2018 article by physicist Paul Halpern in Forbes[11] implies there are natural and stochastic events that greatly overwhelm man's input and what computer models can forecast. For example, nature tends to go to extremes, which brings it back to an average and then the variance back to extremes starts anew.

Here's a classic example: Since 2003, hyperactive hurricane seasons have been followed by large-scale warm winters in the U.S. Conversely, less active hurricane seasons have been followed by much colder U.S. winters. However, snow in the Northeast coastal areas, which is where the focus of warm water and cold air occurs, has been above normal in the hyperactive years *and* the least active years! It's like the extreme opposites lead to a similar result, while more average hurricane years yield less snowfall. One would think that with warmer ocean temperatures, the odds would be for less snow, but even in warm winters you still can get enough opportunity for big snowstorms, and a couple of them tip the scale. On the other hand, if it's a cold and wet winter season, you obviously get good snow.

Watching this over the years makes you realize how small man's influence actually is. The real argument here is how much warming should we attribute to CO_2 and whether the models have

[11] https://www.forbes.com/sites/startswithabang/2018/02/13/chaos-theory-the-butterfly-effect-and-the-computer-glitch-that-started-it-all/#14267d9b69f6

any idea on the true magnitude of it. I have discussed elsewhere my belief in oceanic outsourcing of CO_2, but where is modeling on that? Is it initialized correctly?

What about the effect of hydrothermal vents? Or even Dr. Willie Soon's stunning correlation between total solar irradiance and temperature (left) and water vapor (right)?

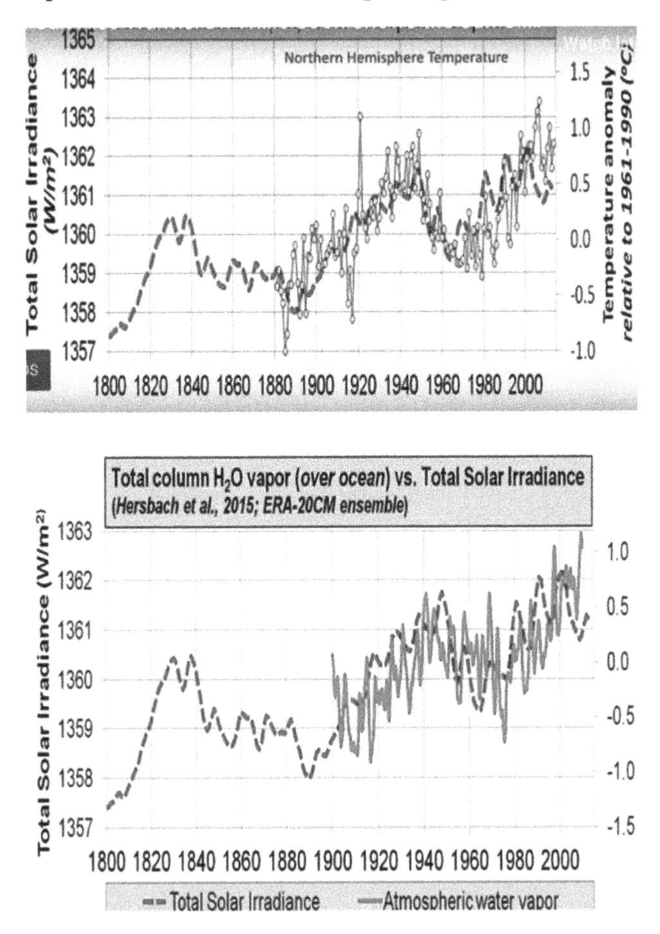

I will never forget how, in 15 minutes, diagramming on a napkin while eating dinner, Dr. William Happer explained to me the diminishing return CO_2 has on temperatures. According to Dr. Happer[12], "Some small fraction of the 1° C warming during the past two centuries must have been due to increasing CO_2, which is indeed a greenhouse gas." However, he adds, "I am persuaded that most of the warming was due to natural causes, about which the governments can do nothing." How can such a brilliant man be fooled? Dr. Happer believes we are pulling out of a CO_2 drought, and more carbon dioxide is better. That really gets people upset. But if Dr. Happer is correct — and I suspect he is — the lion's share of the warming from CO_2 has already occurred, and we are at the hands of nature (and always have been).

Modeling seeks to defy nature with tipping points and runaway warming for three reasons.

1. Too much CO_2 attribution.
2. It flouts the concept that the warmer it gets, the harder it is to make it warmer.
3. Super El Niños and ocean temperatures.

It is absolutely true that, right now, warming is outdueling the cold. Furthermore, over the Arctic, winters are averaging 5°C warmer while summers are neutral. However, I suspect models will not see a decadal change until it's well underway. As long as the oceans remain warm, the air will too. And the outsourcing of CO_2 from the oceans likely has not reached an equilibrium, nor has the biggest greenhouse for climate consideration, water vapor.

---

[12] https://thebestschools.org/special/karoly-happer-dialogue-global-warming/william-happer-interview/

But something may eventually change. I don't think models can forecast the sun's effect over a cumulative term of decades, which has led many to say the sun is not a factor. Shorter term, it might not be. But after a few decades, how can we know for sure? Furthermore, how can a model possibly figure out the effect of shutting off or reducing hydrothermal vents? There is so much of the ocean we know nothing about. And given that the oceans are the source of 99.9% of the thermal heat capacity of the planet, our dearth of knowledge is an important consideration.

Stochastic events, long-term cycles, shorter-term cycles, intersecting then diverging — it's a mad house.

Models are great tools, but they are not the answer. While the increase in CO_2 is occurring as temperatures rise, the temperature *variation* points the finger squarely at natural sourcing. A lot of the policies conjured up by people who seek to impose controls are based on modeling, which runs counter not only to the accepted norms of freedom but of science itself. Then again, this is not about science; it's about curtailing freedom.

CHAPTER 9

The Weaponization of Academia

Schools have become climate-change indoctrination machines. I had an experience with a 12-year-old who was being told that if we didn't stop climate change, we would perish in 10 years. He was assigned a project to show how wildfires are worsening. The teacher was oblivious to the fact that wildfires at the start of the 20th century were much worse. In fact, the record in the 1920s was five times that of 2017. The media capitalized on the recent wildfires in Australia, yet they were about one-third the size of the wildfires in the mid-1970s. But the people pushing a missive don't care, nor are they curious to learn the other side of the story before indoctrinating children.

How bad is it? Imagine taking an Italian language class and the professor tells the students they have 20 years to visit Venice before it goes underwater. Yes, this drivel is included even in language classes.

I wrote a piece several years ago asking how someone who makes climate change their life's work could possibly be objective. It's an interesting contradiction among the pure in heart (there are good people on the other side). One needs relentless focus and motivation to find the answer, which is no different from a champion athlete in pursuit of a medal. However, that same laser focus could also mean you don't see the very thing that could trip you up.

A lot of great leaders believe they can't be wrong or that they can't be a victim of their own success. This makes them dogmatic. And even though their pursuit may be noble, they become intolerant toward other ideas. It's all downhill from there.

What if they're wrong and all the things they taught others over the years was for naught? How bad will that look? Furthermore, if climate change is your one issue — the one you own (which means it owns you) — how can you back off? If you admit to being wrong, it's a betrayal of self and an admission you have betrayed others. The window to recognize you are wrong is very small.

Many of the most brilliant minds in weather and climate refuse to make a forecast that can be pinned down. It's not unlike what I see in bodybuilding. The only way to truly know where you stand is to compete. So you train like crazy (research), and then you compete (make a forecast). I have seen many "gym lifters" who assume they can compete but never do. I see that same tendency among climate doomsayers. They believe their *ideas* are of value, not the *result*.

The value of forecasting (and competing) is that it shows you can be wrong. Just how wrong can be very painful, but once the scars heal, you come out stronger. If you never make a forecast, then you look at weather and climate very differently, and you miss the lessons of humility, open-mindedness, and the fact the atmosphere is a chaotic system. In fact, it reminds me a bit of the speech Rocky gave to his son in "Rocky 6":

> *Let me tell you something you already know. The world ain't all sunshine and rainbows. It's a very mean and nasty place and I don't care how tough you are it will beat you to your knees and keep you there permanently if you let it. You, me, or nobody is gonna hit as hard as life. But it ain't about how hard ya hit. It's about how hard you can get hit and keep moving forward. How much you can take and keep moving forward. That's how winning is done!*

In this case, it's the weather that will beat you down. You never win with nature; you can only adapt and fight back.

Like I said, there are many good people who believe in man-made global warming, and they have grown comfortable in their success, status, and the nobility of their pursuit. What can be more noble than saving mankind from itself, even if it means shutting down other ideas? That said, how do you know if you're simply moving with the masses (the "consensus")? You can't know for sure unless you're constantly testing yourself and actually studying and listening to the other side. The ideas you should be most skeptical of are *your own*.

Be it making a forecast or putting extra weight on a squat rack, you must be willing to get in the arena and take the hits.

I have a great deal of sympathy for these people, because in many ways I'm just like them (just not with how I view the climate). But I realize I might be wrong. It's not my way or the highway. Then again, my whole being doesn't depend on me being right on this issue. If I'm wrong, it's not the end of the world. Man has always adapted to challenges. This perspective is a far cry from the idea you *can't* be wrong and the belief that the fate of mankind rests on your shoulders.

There have been many times I was *convinced* of how a weather event would play out. That's led to a lot of great standout forecasts, but also some busts. The focus has to be there. But what also has to be there is the willingness to do whatever it takes to come up with the right answer, even if it's not the answer you wanted or expected.

Academic indoctrination is also driven by money, greed, and ideology. These make up the really dark side — the seduction of it all. All three motives are fatal flaws to the objective pursuit of knowledge. Ideology is intolerant of other people's opinions. Money is used as a weapon to shut down other ideas. Greed takes over as one becomes entrenched in it. For example, to get a research grant, you have to come up with ideas that back the ideological-like global-warming missive. Perhaps this isn't so much greed as it is survival. Almost a case of indentured servitude.

I feel like Dante Alighieri describing layers of hell here. We started with the good-intentioned and very smart and then dropped into a more devious motivation. Finally, we reached the lowest of levels — the eradication of this nation's foundations. The latter is what this is really about. Each of these chapters is an attempt to get the reader to understand that, in a way, Dr. William Gray's claim that the field has become riddled with climate pimps who are prostituting weather and climate for their own purpose was right. This last level of hell relates to journalists, who mindlessly write what they're being told, and politicos, who make statements that run contrary to facts (and journalist never call them out on it). Journalists and politicians are mouthpieces in a cottage industry who monitor and attack anything that may run contrary to their mission.

The scariest thing? They have already taken control of several generations. And it's only a matter of time before that chicken comes home to roost. You can always vote your way into communism, but you have to shoot your way out.

I'll close with an observation from Albert Einstein: "By academic freedom I understand the right to search for truth

and to publish and teach what one holds to be true. This right implies also a duty; one must not conceal any part of what one has recognized to be true."

CHAPTER 10

COVID-19/Climate-Change Alliance Weaponization

When I began writing this book, coronavirus was in no way meant to be in it. However, I felt that it had to be mentioned after witnessing what I did.

As predictable as the morning sunrise, we are hearing that the spread of COVID-19 is enhanced by "climate change." Not only is there no proof of that, but it runs counter to the fact that life *thrived* in warmer times. As we've already established, they were called optimums because life thrives when it's warmer.

One could argue that a virus is a living organism, but the point is that if viruses thrived in the past when it was warm, plant and animal life — which includes humans — thrived more.

Over the years, I've made no secret that I believe the climate-change agenda is a smoke screen for other agendas. One of them is driven by the idea there are too many resource-depleting people on the planet. This flies in the face of God's command to "be fruitful and multiply," which, of course, doesn't mean be stupid and trash the planet. Then again, it also doesn't imply there is a set limit on what man can do with free will and a head turned toward a higher calling. Yet when I look at some of the statements put out by prominent people hitched to the rhetorical climate-change bandwagon, I realize there's a strong connection to population *control,* the idea being that one person knows better than another person what's good for society as a whole. Like it or not, Socialism (or Marxism) isn't about equality. It's a top-down form of government control in which the vast majority of the people don't have a chance to rise into the upper echelon unless they pledge loyalty to that doctrine. In essence, it's the destruction of free will. This runs counter to God's gift of free will.

It comes down to *you* making choices or *someone else* making choices for you. Where's the evidence for this? I think a few quotes will make my point.

- "A total population of 250-300 million people, a 95% decline from present levels, would be ideal." —Ted Turner
- "A cancer is an uncontrolled multiplication of cells; the population explosion is an uncontrolled multiplication of people. We must shift our efforts from the treatment of the symptoms to the cutting out of the cancer." —Paul R. Ehrlich, President of the Stanford University Center for Conservation Biology and Bing Professor of Population Studies
- "There exists ample authority under which population growth could be regulated... It has been concluded that compulsory population-control laws, even including laws requiring compulsory abortion, could be sustained under the existing Constitution if the population crisis became sufficiently severe to endanger the society." —John Holdren, President Barack Obama's science czar
- "The extinction of the human species may not only be inevitable but a good thing." —Christopher Manes, a writer for *Earth First!* journal
- "My three main goals would be to reduce human population to about 100 million worldwide, destroy the industrial infrastructure and see wilderness, with its full complement of species, returning throughout the world." —David Foreman, co-founder of *Earth First!*

- "Childbearing should be a punishable crime against society, unless the parents hold a government license. All potential parents should be required to use contraceptive chemicals, the government issuing antidotes to citizens chosen for childbearing." —the late David Brower, executive director and board member at the Sierra Club
- "The resultant ideal sustainable population is hence more than 500 million people but less than one billion." —Club of Rome
- "For the planet's sake, I hope we have bird flu or some other thing that will reduce the population, because otherwise we're doomed." —Susan Blakemore, *UK Guardian* science journalist

An exhaustive list of quotes can be found at *C3 Headlines*[13]. There is more to this issue than just a temperature increase of 1°C.

I've also seen the claim that climate change is worse than COVID-19. This is yet more proof that the "climate change" issue is more than what meets the eye. It's about using any tactic possible to push an agenda that's regressive and seeks to *limit* what man can do. For example, a March 15 article in the UK's *The Times*, entitled "Climate change will be deadlier than Covid-19,"[14] implies something not supported by any facts today to scare the daylights out of people who are already on edge. We are all anxious about this virus, and yet here's the predictable use of climate change to scare people even more.

---

[13] https://www.c3headlines.com/global-warming-quotes-climate-change-quotes.html

[14] https://www.thetimes.co.uk/article/sarah-mcinerney-climate-change-will-be-deadlier-than-covid-19-lstrdl6t7

The facts are clear on climate: Over the last 100 years, the impacts on mankind are *exactly opposite* of what was predicted. Which is even more remarkable given the increase in human population.

These scaremongering journalists have neither the intellectual curiosity nor the integrity to do what journalists are supposed to do: Question and show both sides of the issue.

In direct defiance of what's implied in the above article, the truth is that climate deaths are plummeting. Globally, personal GDP and life expectancy are skyrocketing. This means more people are living longer, more prosperous lives. GDP and life expectancy began trending up at the start of the fossil-fuel era. More food is being grown than ever before. The earth is greener than ever before.

As an aside, consider this (and don't think this is minimizing the COVID-19 threat): The median age of deaths is in the low 80s. But if this virus appeared in the pre-fossil-fuel era, it may have gone virtually unnoticed because life expectancy in the pre-fossil-fuel era was between 30 and 40 years[15].

Be on guard against any event being blamed on "climate change," because it's really a vehicle for another agenda. At the very least, consider the actual data presented here.

My advice to journalists: Look at both sides of the issue. You might come to realize that the foundation on which you stand today was built during the fossil-fuel era. Don't take that for granted! Be open-minded, be tolerant, and don't accuse people who bring up ideas you may not have considered before of being a science denier.

---

[15] https://ourworldindata.org/life-expectancy#life-expectancy-has-improved-globally

The False Equivalency Between Climate Change & COVID-19

CHAPTER 11

This chapter is not meant to minimize the threat from COVID-19. It's meant to point out similarities in the climate and coronavirus messaging.

The end result of COVID-19 is unknown. That means predictions have a wide range. Believing and *knowing* are two different things. We see issues arise in the fight over the effect of CO_2. People like me believe that as CO_2 increases, there's a diminishing return in the atmosphere. Others disagree. The corollary arguments range from nothing at all to limited effect to runaway warming with virtually no limit.

Now consider the ranges you're seeing with COVID-19, which run the gamut from no discernible effect to regular expectations of mortality with age to the kind of numbers that would rival major past epidemics. Lost in all this is the fact we have far more people living in the world today, meaning mortality percentages are lower. But given our more prosperous world today, the effect on the quality of life is indeed devastating. Additionally, with a bigger global population, more people are going to be exposed and perish. That's a function of the median age rising and the explosion of population and prosperity. It's extremely serious.

* * *

I have some close personal friends in the medical profession. Many of them are looking at COVID-19 as yet another disease to which we must adapt. There are even a few who believe the response of "an ounce of prevention is worth a pound of cure" is worth it even if this turns out to be a best-case scenario. They realize that until the unknowns become knowns, it's wise to prepare.

However, in the media, good news, or even potentially good news, is being either ridiculed or not reported. We've morphed into a situation in which you have both people of goodwill and people who desire a set outcome. In other words, an agenda has developed. And once a person becomes invested in a result, that person's vision can become clouded to anything that might tarnish the desired outcome.

As someone who sometimes sticks too long to collapsing forecasts, believe me, I'm not trying to be critical. But I see a common denominator. When someone is saying that climate change is worse than the COVID-19 outbreak, that should raise red flags simply due to the timescale. As we've witnessed, COVID-19 can tear down an economy in mere weeks. But before that, the U.S. economy likely did better than was expected last fall due to the mild winter. That's obviously a positive result — unless you're against progress.

The point is that there are very good people who disagree, but at the same time, an agenda-driven cottage industry has developed.

One test I've employed is detecting tone. For example, take the anti-malaria drug hydroxychloroquine. The possibilities of what can go wrong are no different than what you see on TV for other preventive drugs, which come with a whole slew of possible side effects. I've read the studies regarding hydroxychloroquine. And even though it isn't the holy grail or a magic bullet, there *are* encouraging possibilities. Yet the tone from those trying to throw cold water on the drug is ugly. And if that tone were genuine, it would be applied to all common drugs on the market.

Watch a TV ad for any drug. Sometimes one of the side effects is death. I think, "Why would anyone take that?" The idea that there is reward without risk is absurd. And blasting the president for creating hope or trying to lead is a closed-minded and intolerant approach. The objective person wants coronavirus answers, while another whole subset of people is working to instill an agenda.

* * *

Another similarity to the climate-change debate has to do with how the issue is approached. Racist claims should raise red flags. *The Federalist*[16] has compiled a list of more than a dozen diseases whose names were derived from populations or places. A few examples include West Nile Virus, Guinea Worm, Lyme Disease, Ebola, and Spanish Flu.

Should all of these diseases be "erased"? In the same vein, if you want to argue that the current warm period is an emergency, should we erase even warmer periods in the past that were listed as optimums?

* * *

Forecasts of doomsday are as old as mankind itself. A recent article published at CFACT[17] revealed how wrong forecasts were

---

16 https://thefederalist.com/2020/03/13/17-diseases-named-after-places-or-people/

17 https://www.cfact.org/2020/03/13/greta-preaches-many-of-the-first-earth-days-failed-predictions/

from the very first Earth Day. Many of those same forecasts are showing up now, but of course for a different reason. Only time will tell if the worst forecasts will come to pass. But in almost every case, the predictions have not become reality, which is why they are never brought up.

* * *

I believe in less government, but we do need an essential framework. There's an argument in the climate doomsayer community that we should have the same kind of response to climate that we have for COVID-19, which is interesting in that, if anything, COVID-19 is showing you what happens when the business engine of the nation shuts down. It leads to misery. But no one can argue that before the virus hit, life on the planet wasn't much improved over the last 100 years despite — rather, *because of* — global warming. The coronavirus shows us that government should step in for legitimate risks. But it's false to suggest that a more gradual event like climate change is just as legitimate a risk.

You're seeing what taking trillions of dollars out of the economy can do. Now imagine a $10 trillion plan to fight "climate change."

* * *

I'm seeing alarmist tactics used to ramp up coronavirus fear. Consider this: In examining the Johns Hopkins COVID-19 dashboard for March 29, I found that the curves for the U.S. and Australia were rising in almost identical fashion. But then

I checked the scale. The surge of 20,000 cases in the U.S. dwarfed the 400 in Australia. In other words, by superimposing the Australian figures on the U.S. scale, I discovered that Australia's numbers would barely move off the straight line in the States.

Yet with the exception of South Korea and China (the latter's figures are suspect), countries with a far lesser increase in cases are better represented. This is similar to making minute increases in temperatures look bigger by adjusting the scale, leading to a wicked distortion. The same goes for CO_2. All scales have to be the same in order to see the big picture.

Another area I see parallels between climate change and COVID-19 is in the lack of perspective.

The true measurement of this disease comes via computing how many people are dying daily versus the average for the date, season, etc. Or how many people are dying with no preexisting conditions versus the average. Yet no one dares to bring that up. And until the whole thing plays out, we won't know.

Climate-change extremists immediately climbed on board the COVID-19 bandwagon. The first volley was to accuse the global climate "emergency" of being complicit in the spread of the disease. This was distasteful on two fronts. First, it showed that these people will use anything to push their point across, revealing a dangerous zealotry that points to even harsher tactics. Second, they failed to realize that most viruses don't do as well in warmer conditions. The postmortem on this disease will be to determine what if any effect the weather has. It may be like other flu strains in that it weakens in the summer and then comes back in the fall.

* * *

On social media and in the mainstream media, the level of viciousness is mimicking the kind of vitriol I see in the climate-change debate. I've been accused of minimizing the threat of COVID-19. The opposite is true. I am 64 years old, and many of the people I cherish the most are right in the middle of the age demographic this disease is targeting. I had a dear friend of mine say he wanted to ensure I wasn't cavalier about this virus. For me, it's a test of faith. What concerns me is the derisive branding of people who offer different ideas, even people whom you may agree with on almost everything else but who get upset because you don't see this particular issue the same way they do. And they will do anything to force — keyword being *force* — another person to bend to their will.

Attacking people is an attempt to censor ideas. It's close-minded and intolerant. You don't learn anything by shuttering ideas. I welcome discussion. If I'm wrong, it makes me stronger because I learned something.

Obviously, this is a very serious situation. One of my professors at Penn State used to say that advisories for snow can cause more disruption than the snow itself. (For those of you in the South, how many times have you seen snow in the forecast and the shelves laid bare, even though the snow is gone after just two days?) That possibility exists with COVID-19.

I believe that the pain we are going through will be worth it, even if the virus fizzles out completely. I can't say the same about using this as a strategy to develop a war footing on "climate change." Freedom is needed to pursue the right answer, and those wishing to limit that freedom by demonizing and destroying those with whom they disagree are motivated by a *desired* outcome.

Please stay safe and heed the cautions. I don't know or pretend to know the outcome of COVID-19, even though I may believe in a certain result. But just like with forecasting the weather, I can't let contrived notions cloud my vision. But one thing is unequivocally true: Open-mindedness and freedom should be used as keystones to face challenges.

Using Climate Weaponization Against Freedom

CHAPTER 12

I wrestled at Penn State under a guy named Bill Koll. Next to my dad, he was the most important male inspiration in my life.

Who was Bill Koll? Here's a snapshot courtesy of the University of Northern Iowa[18]:

> *After World War II, in which he participated in the Battle of the Bulge and D-Day, Koll dominated the 145 pound and 147.5 pound weight divisions. In 1946, 1947, and 1948, he was the NCAA national champion for his division. Koll was also named the NCAA Tournament Outstanding Wrestler in 1947 and 1948. He ended his career with a 72-0 record. Koll was one of only five UNI wrestlers to have competed in the Olympic games. In the 1948 games, held in London, England, Koll placed fifth.*

Coach Koll used to say that wrestling was America's sport. I excerpt myself here from a January 2010 article at StateCollege.com[19].

* * *

I took Coach Koll's wrestling coaching class as an elective. And one day, he explained why wrestling was truly America's sport. I wish I had a tape of it.

---

18 https://scua.library.uni.edu/university-archives/biographies/william-koll

19 http://www.statecollege.com/news/columns/a-reading-from-the-book-of-koll-why-wrestling-is-americas-sport,309182/

America was the only country that understood the idea of freedom leading to excellence, he said. But the main lesson was this: The rise of the individual leads to the rise of those around him. This assumes, of course, that those around him want to win and have the freedom to compete.

The team is the Constitution and the coach is the president because he sets the tone.

Here is how he would explain it:

> Person A lives in a town and makes widgets. He employed only one worker at first but employs dozens now that he is the Widget King. (Back in those days, they used to talk about widgets all of the time.)

The whole town is benefitting from the Widget King's success. People who want to be employed are employed, and he makes a profit.

On the other side of town is another guy who thinks he can make widgets better. So he challenges the Widget King. Pretty soon he is employing more people, and the quality of the widget improves. Now the whole town is the widget capital of the world.

But what if someone said they could not make those widgets, or made it so they could not turn a profit off the widget? Why then would they do anything? (Here he would point out the problem with communism). It took away the incentive to excel.

He then tied it into the team. As "president," he had a certain philosophy that said we would be in better shape than anyone and take the match to our opponents.

He would show us the moves, but we had to perfect it. We had to do the work on our own if we wanted to excel. In addition, we would find ways to butt heads against the guys you wanted to be better than. We had the freedom to train ourselves.

But he would tie all that in and make a civics lesson out of it.

If it was good enough for a three-time national champion, an Olympian and one of the first men on the beach in Normandy, it was good enough for me.

Koll always felt that a winner wanted to be challenged. My first two years on the team he hardly said boo to me. (One time he asked me if I liked basketball because I looked like one the way I was getting bounced up and down off the mat.) But my senior year, he rode me without mercy.

He was always a nice guy in the office, so I went in and asked him what was going on. He said, "Joe, you weren't worth anything before, but you worked so darn hard to get where you are, I figured I should help you."

Within the realm of his "country," I used the freedom he gave me to put myself in a position where he would pay attention. He was going to help in his way.

It was civics lesson that no college course could possibly teach.

* * *

Coach Koll believed that someone who truly wanted to win always endeavored to correct their weaknesses, not be patted on the butt for their strengths. It's much harder to go from 99-0 to 100-0 than to go from 0-0 to 1-0.

I find it fascinating that when I talk to my former teammates about a lot of the lessons Bill taught as a coach, they don't really remember them. It hit me that I was so bad at wrestling that I hung on to every word Bill and my assistant coach, Andy Matter, said. But if you are really, really good, you probably just think you're being told what you already know. Coach Koll used to say some guys could have been better if not for their stubbornness.

I call it the curse of talent — the strong belief that you are so good that no power can help you other than your own.

What does this have to do with the climate? Key aspects of the curse of talent run rampant in the climate-alarmist community. I know it well because I saw it in wrestling *and in myself.* When I was at PSU, I think it was generally acknowledged I was the leader of the weather junkies. I had my own forecasting radio network by the time I was a junior. I had a job offer at the end of my junior year to work for AccuWeather, which I believed at the time was the New York Yankees of weather. Tremendous talent under one roof, passionate guys who would fight over a forecast five days away over a low temperature. In other words, people just as nuts as I was.

But I thought I was as good or better than the veteran forecasters there. Dumb move, because I shut out a lot of opportunities to grow. If I had simply not tried to show I was better than them, I would have learned even more earlier in my career and become even better. It was as if I could hit a fastball (monster weather events) but strike out on changeups (small-scale weather patterns). When I was in the wrestling room at PSU, I *knew* everyone was better than me, so I had to pay attention, work harder, and be open to new things. I knew so-and-so could beat me on the mat, but I also believed I would go further than them professionally because of the talent I had in meteorology. That was wrong also. Almost all of my teammates are very successful at what they do. They learned the lessons and applied it to their post-college careers.

A lot of us feel we owe our success to having wrestled at PSU. Once they were confronted with a challenge, my teammates responded by doing in their jobs what I did in the room to improve — listen and learn. The key is to never believe you are the best. Instead, give thanks for the situation you are in and work to be the best.

After graduating from college, I came to recall what my dad taught me: Compete against reality, not people. My attitude eventually reverted back to what it was when I entered PSU as a freshman. Except now I focus not on what individuals are doing but on the majesty of God's creation. To get better as a person, I must be open to other viewpoints. I'm the dumbest guy in the room with a desire to find the answer.

This long-winded discussion sets us up for what I see as the weaponizing of climate against freedom. A great wrestler (and forecaster) wants to be challenged and lives to see how he or she performs in uncharted territories. Which is why I look much more at what people on the other side of the climate debate say than I do at what I believe. I already know what I think; I want to know what *they* think. That's how you get stronger. If you travel the same road every day, the scenery never changes.

Glance through any article portraying a notable weather event as out of the ordinary. What reference will you see every single time? Something along the lines of, "It's due to climate change." Part of me understands why. If you sink 30 years of your life into an agenda, there's no way you turn back now. Yet this is completely opposite of how to improve as a person. And here's where this comes full circle: People like me relish freedom, challenges, and competition. The Founding Fathers knew that these three ideals would drive progress. I believe in a mighty and merciful God and that all people are created equal but with different talents. It's freedom, challenge, and competition that enable individuals to reach upward mobility. Yet we're seeing the opposite happening.

For starters, our children are being indoctrinated by a system that says *they* are the "authorities." I get a kick out of that. How can you have authority over something you can't control? Unless, of course, you're convinced that you *do* have control. I believe God is in control, and humans advance by using freedoms given to us to react and adapt to circumstances inherent in the system. Contrast that to the other side. Inherent in their belief system is a "holier than thou" attitude that either denies a higher calling or substitutes it. This requires the shutting down of free thought and trials. And what better place is there to do that than in the climate arena?

To accept the idea that humans can save us, you have to be indoctrinated. And indoctrination destroys freedom. I do have compassion for those on the other side of this issue. Many of them are very smart and well-intentioned. But the road to hell is paved with good intentions. The very thing that can expose them to the truth is something they will not dig into because they would see the climate for what it is — a question that is not simply answered with tropes. And that's something they don't want to face.

Regardless of whether they are ultimately proven right, their *tactics* are far more damaging than temperatures.

CHAPTER 13

Political Weaponization

Look at the major issues of the day. They all have one thing in common: While the issues deserve debate, one side wants all dissension shut down and demands blind faith.

Climate-change rhetoric has reached campaign-ad levels. Talk of declaring "climate emergencies" and "saving the world" is appearing. There is a tidal wave of this coming in the upcoming election cycle.

Most of the people on my side of the issue know and understand that CO_2 is in the equation, but they question what it can ultimately do given all the other natural drivers. We also understand there is valuable research and points being made about CO_2, but they need to be scrutinized. We love to debate and engage. But we aren't being asked to question — just obey. That's a foreign concept to old fogies like who me who grew up being taught to do the opposite.

Believing and knowing are two different things. You can believe something will happen, but until and unless it does, it's only speculation. You can *believe* there's a climate emergency, but competing evidence suggests you can't actually *know* that.

There have been numerous climate optimums throughout history. How does this impending one suddenly constitute an emergency? You would have to rewrite and/or deny history to say this is an emergency. What we're being asked to believe is that, in order to save the world, we have to mobilize on a war footing to prevent a climate emergency. Does that make sense?

Furthermore, previous temperature fluctuations were *not* caused by humans. Why is this one any different? Isn't it far more likely that the cause today is the same as it was before?

A friend of mine, Gregory Wrightstone, a well-known author and geologist, recently said: "The warming effect of each molecule of carbon dioxide decreases logarithmically with increasing concentration. And THAT … is why there was no 'runaway global warming' in the past when CO_2 exceeded 5,000 ppm."

Lord Chris Monckton has used the Intergovernmental Panel on Climate Change's own formula to show the diminishing return on temperatures when CO_2 increases. Dr. Will Happer has been trying to get that point across also. The fact is, the overall global temperature increase is not 4-5°C. It is more likely capped at about 1.2°C. If so, man's influence is nearing a crescendo. Moreover, if global warming is entirely the result of CO_2, it's done much of what it can do anyway. Since plants grow best at around 1,500 ppm, perhaps we are heading toward an equilibrium.

However, the tricky question to the lockstep left is this: How do you reconcile the idea of a climate emergency when we are nearing a climate optimum? My climate-change equation includes the effect of the sun, the cyclical nature of the ocean, stochastic events, and the very design of the system. Most of the stored energy (warmth) is in the oceans, two-thirds of which are located in the southern hemisphere. This creates a major natural imbalance with the land in the northern hemisphere, where the North Pole contains an ocean. This is unlike the South Pole, which sits on a continent. The planet is also rotating elliptically around a star (the sun) that varies in its total solar irradiance. All of this *far outweighs* the effect of $CO_{2.}$ Therefore, it seems intuitive that the cause is mainly natural.

The Obama administration's EPA said that the same restrictions that are largely included in the Green New Deal come at a price tag of $3-10 trillion — all to save .03°C over 100 years.

Ostensibly, that would show the rest of the world we are serious. It's hard enough to get a trade deal with China. Are these people willing to go to war with nations that won't follow our example? What if other nations, now stronger than a weakened U.S. due to the draconian steps taken to combat the climate "emergency," don't want to obey the suggested course of action? A quote from H. L. Mencken, a Democrat, comes to mind:

> *The whole aim of practical politics is to keep the populace alarmed (and hence clamorous to be led to safety) by menacing it with an endless series of hobgoblins, all of them imaginary. … The urge to save humanity is almost always a false front for the urge to rule.*

Is the climate agenda not the very example of what Mencken was alluding to so many years ago?

My take is that someone could ride this issue all the way to the top. Laugh now, but as stated before, the America First crowd has to come up with some plan that will neutralize the expanding bloc of young voters. People who find "saving the planet" important or having their student loans forgiven will vote for the candidate who echoes them without thinking about the consequences. When you're not taught to question the issues, all you do is obey.

There has to be something that at least appeals to the sense of objectivity, humility, and productivity for our nation going forward. Otherwise, future generations will be severely impacted by Socialism/Marxism. Thirty years ago, young people fought like tigers to get rid of the Eastern Bloc. It strikes me as ironic that today's movements are trying to do the opposite.

Many people blast Fox News because of its strong opinion lineup, but Tucker Carlson, Sean Hannity, and Laura Ingraham love having people debate them. Yet for all their individual prowess, they are stacked against more networks that simply push the agenda du jour down the throat of the American people. You can take any issue out there and find the same kind of attitude. Climate is but one aspect, but it's out there front and center.

I don't think there's any reason for this to be a big issue given the planet's history. But for doomsayers, it's a means to an end — a weapon geared toward turning the nation into an Eastern Bloc-type dystopia. And because it's political, it's not going away. Therefore, it must be questioned and responded to.

Wildfires

There are many reasons why some wildfire seasons are worse than others. The last three spring seasons I have made predictions for big wildfire seasons based on a big factor — winter and spring rainfall. The more it rains in the winter and spring, the more foliage there will be. And because California is normally dry in the summer, that foliage is going to add extra fuel to wildfires. This is by no means the only reason, but it helps tip the scale. And it comes as no surprise. It's akin to saying there are hurricanes in the hurricane season. Guess what? Most years someone is going to get hit. The years in which there's abnormally low activity and damage are the exception, not the rule — especially with more people living in harm's way.

It's tough writing about this because my heart truly goes out to those who are being devastated by these events. They become pawns in this game. The compassion in almost all of us tugs at our heartstrings. It makes anyone who brings up facts for perspective seem heartless and cruel. Yet we must deal with the reality.

There are three times as many people living in California today than there were in the 1950s. This means more people are in harm's way. More people in the woods mistakenly starting fires and power-gird problems are enhanced by the insistence that we should not be clearing out more deadwood.

One supposed culprit cited by the media — which has become a willing accomplice in making sure this issue is one-sided — is climate change. And it's being pushed further by the people in charge in an effort to advance their agenda.

But here's the reality: Since 1980, total acreage burned has indeed been increasing in the U.S. But thank goodness it's not the old normal. If we look at the entire picture, the total acreage burned in the 1920s to 1950s was *far* higher despite a lower population. Climate-change manipulators like to compare temperatures today to 100 years ago — so why not compare wildfires? Maybe they don't want people seeing the big picture.

Wildfires are another case study in weaponizing weather events to push an agenda without showing the other side of the argument. It's like this across the board now. There are reasons to question things now more so than ever, but you have to *look*, not just accept what you're being told.

Remember also that when it comes to devastating events like wildfires, we are far better off now than we were 100 years ago. Fossil fuels have played a big part in the advancement of mankind. *We do not own nature.* We reside and advance here *despite* nature. But the idea that man can create a utopian Garden of Eden represents an exercise in arrogance.

Is there more acreage being burned than in 1980? Yes. But there are many reasons other than climate change, and it's not nearly as bad as it was 100 years ago.

By the way, in Australia, there was a wildfire season in the 1970s that burned an area the size of France, Portugal, and Spain — far greater than this past season, which burned an area the size of Switzerland. It's just that now there are more people living in the area, yet historical perspective isn't even brought up. The weaponization-of-the-weather agenda requires hiding past events that can diminish the severity of what's portrayed.

Sea-Level Rise & Hurricanes

If you're going to weaponize sea-level rise, don't buy a house on the beach. Before I get to the reason hurricanes are not as bad as they could be — at least not in this age of warming — a great confirmation of the weaponization of a phony threat is when actions don't match words, as in the case of our former president.

If President Barack Obama had asked me to be part of a team on climate change, I would have jumped at the opportunity. That's the way I was raised. If your president calls you, you respond. I would have been the only dissenting voice, but I would have participated by making my points and showing what I believed. My point is that I have no qualms about mixing it up with people who don't agree with me, and I don't hold their ideas against them. However, when it comes to *action*, that's a different story.

I have trained a lot of great athletes — from Olympic heavyweight Kerry McCoy to former New England Patriot Sean Mayer. The one thing I never did was put them through something *I did not do myself.* In truth, I also trained *with* them, because if some old man is going hard, the trainee will go harder. I try to surround myself with people who are smarter or stronger than me. This yearning to get what they have often creates a mutual feedback.

President Obama, by purchasing a mansion on Martha's Vineyard, demonstrated hypocrisy. It proved that his stance on climate change and rising sea levels can't be taken seriously because he's acting against what he claims he knows. But he's apparently also unaware of the prodigious hurricanes that affected the New England coast in 1938 and 1954, when CO_2 levels were far lower than they are today.

I suspect many of his friends in The Hamptons are similarly unaware. Otherwise, why would anyone build there when the 1938 hurricane was accompanied by a storm surge that flattened the east end of Long Island? If you want to build and live on the coast, it's not a changing climate you need to be afraid of; it's the inherent risk of nature.

In 1954, Hurricane Carol came a-calling. It formed north of the Bahamas and eventually made landfall on Long Island on August 31. Hurricane Edna followed suit. This one originated north of Puerto Rico and moved north until heavily impacting Long Island again just 11 days later. These back-to-back hurricanes occurred when CO_2 was far lower. So why *shouldn't* it happen again?

Hopefully you understand why it's baffling to me that anyone would purchase along the coast. I think the former president should live in a great house — the biggest one he can get. I just don't want him telling all of us how bad the climate is and then making personal decisions that contradict his concern. Is that leadership? No. It's politics. Don't fool yourself. There's a death warrant out for capitalism, and climate-change dogma is a means to an end.

In contrast, I imagine President Donald Trump knows the inherent risks involved with owning Mar-a-Lago. He comes from a business background, so he indubitably administered a risk-benefit analysis before purchasing it. The Mar-a-Lago area was Grand Central Station for major hurricanes in the 1940s. From 1944-1950, *seven* major hurricanes hit Florida — primarily the bottom half. A repeat would cause innumerably more devastation today.

Since the day President Trump took office, two things have become apparent. 1. There are high-profile people who want him out of office as fast as possible by non-electoral means. 2. There is a desire for his southern White House to get hit by a hurricane to make a point about climate change.

These people know darn well the general public is unaware of how bad it was in Florida during the 1940s, when CO_2 was far lower. They also seem to ignore the fact there's going to be a price to pay when erecting a lot of buildings and pouring a lot of concrete in hurricane-prone area. At least President Trump acts the way one would expect. I would *anticipate* him to occupy a plush spot on the coast. That's who he is. But if President Obama is who he says he is, why would he buy a house in an area with a history of major hurricane strikes? President Obama's actions boil down to, *Do as I say, not as I do.*

Here's the problem: There are many voters, particularly young ones, who won't look into this issue. You're not going to be able to convince them that if a major hurricane hits, it's not evidence of climate change. So advocates of life, liberty, and the pursuit of happiness *need* to come up with a plan to neutralize the left's hold on this issue.

There *is* a rising sea — that is, a burgeoning climate-change voter bloc. I'm not a politician, but I do understand that if a political party is to continue winning elections, something must be done to confront this problem. Something that won't hurt our economy but will be a fail-safe position.

Unless the field changes, people who are willing to spend $10 trillion to (allegedly) save .01°C over 30 years will come to power. This will spell the end of our economy and prosperity.

De-Weaponizing the Real Battlefield

CHAPTER

The real battleground is being obscured by the smoke of the climate-change agenda. There are people on both sides of the debate who are truly invested. But if there was any kind of even playing field, you would know the names of prominent climate skeptics. They are numerous, but they are also isolated, demonized, and destroyed. They get branded. The dismissal of people whose PhDs required just as much knowledge as did the doctorates obtained by their peers with opposing viewpoints should instantly make any objective person dig deeper.

The constant bombardment on young minds with the idea they don't have a future is something that can't be overcome. Your intentions may be noble, but the horse is out of the barn. The left owns this issue. That voting bloc will be crucial in coming elections — enough to tip the scale. If that happens, it allows socialism to carry the day, and Ronald Reagan's warning that we can lose our freedoms in one generation will have been proven true.

The solution? *Make this our issue.* Present a plan that won't crash the economy. Put forward ideas that will enhance the future of that voter bloc and capture its imagination. Let these voters know we aren't some exclusionary greedy machine but one that truly loves our neighbors and wants the best for them. Explain the humility that comes from understanding the challenges we face. Admit you're concerned about the real cause of global warming. Take CO_2 off the table by implementing the following three-pronged plan I call the Real Green Deal.

1. Plant trees. In fact, create a global Marshall Plan to expand warm-season foliage. Studies show that planting enough trees around the globe — enough to fill Canada — would take care of the CO_2 increase, no matter the cause. The U.S. could take the lead. It would instantly get the attention of people who think we are only about bulldozing and burning trees. It would be far less costly than a $10 trillion (or more) Green New Deal that's nothing more than a smoke screen for the redistribution of wealth. What's the other side going to do — be against planting trees? That response would expose these extremists for what they are.
2. Embrace nuclear power. It's time we shake off the China Syndrome hangover in this country. This is the 21st century, and we are the United States. We can build safe nuclear power plants, and we can figure out a way to get rid of nuclear waste. This is something even James Hansen — one of the prime drivers of the man-made global-warming narrative — agrees with. It's a CO_2-neutral way of generating power.
3. Implement carbon capture. This would not only significantly reduce the release of CO_2 into the air but also protect jobs in the oil and coal fields while opening up an opportunity for more jobs. This would have the added benefit of safeguarding fossil fuels — the most efficient source of energy we have.

Item No. 4 is the string that ties the package. I'm involved with a group that has put together a brilliant economic plan that would not only prevent shutting down our economic lifeline but would put the choice into the hands of the consumer. By the time you're reading this, the details may or may not be in the public arena. But the plan puts America first and, just as importantly, puts *you*, the consumer, first. It gives you a choice in how to handle this matter.

CO_2 is being used as a tactical weapon that's not only armed to the teeth but has created a phony war. What's the best way to counter that? Take the issue out of the protagonists' hands with ideas that will make CO_2 a moot point. If we develop a perfect balance between CO_2 emitted and CO_2 done away with, yet the cumulative number is still climbing, then humans can't be at fault. Either way, the issue is off the table.

It's very important that freedom-loving people develop a bold idea to capture this crucial set of voters. You aren't going to convince them with scientific arguments since most of them have been brainwashed all their lives and really don't have the time and/or inclination to look into the counterevidence that our schools, media, socialites, and politicians have bottled up. Instead, prove that it's our side — not the side jamming rhetoric down our throats — that has the humility to say we're concerned and want to provide options. It's our side that wants to put forward true solutions.

We can de-weaponize the issue and the result will be a positive one. The opposite is true if we capitulate to zealots.

Any solution to any problem should have to answer this question: Does it do harm to life, liberty, and the pursuit of happiness? The plan outlined in this chapter does no harm and, in my opinion, builds on America's foundation.

Conclusion

I have presented a case in this book for why climate change is a smoke screen for a political, social, economic, and even anti-religious agenda (a phony war). It's an offshoot of the cultural, political, and spiritual upheaval that's happening. Trying to elevate what should be a debatable subject to "war" status is an attempt at making yourself a hero. In essence, instead of one nation *under* God, we may become one government *instead* of God.

Which would take us to Marxism — the elevation of the state to a point where the person in charge gets to determine what the social good is at the expense of the individual. This book is not meant to downplay the valuable research of the people who are genuinely concerned that the planet is careening toward a climate meltdown. They have noble reasons, and I have said many times I would believe a meltdown is coming too if CO_2 was the only thing I took into account.

You won't gain knowledge by simply looking at what you already know, so I can't dismiss the deep research being done. However, I get the impression that many people are not intimately involved with day-to-day weather and what it's capable of doing. This book is meant to show curious minds that climate change is not the issue they think it is.

In the song "Leningrad," Billy Joel writes: "Haven't they heard we won the war/What do they keep on fighting for?"

The real fight is over freedom. Climate is being used as a vehicle to control the masses. People don't overthrow governments that are helping to make their lives better, and a warmer world, no matter what the cause, would do just that. We are sliding into a society that runs counter to our foundational values.

The elevation of authority to the point where the individual is dependent on the state for survival diminishes life, liberty, and the pursuit of happiness. If you don't believe in God, that's all poppycock to you anyway. But consider this: You are not in control of anything, no matter what you believe. You may think you have control, but you will go out the same way you came in — helpless. And a helpless crowd will naturally turn to state authority to create a certain social order.

I have stated my beliefs, but you don't have to accept them. Therein lies the difference. A person who believes in God's will and His greatest gift — free will — will naturally recoil at being forced into a situation where free will is taken away. So the elevation of government over God is very much at play here — at least behind the scenes.

Ecclesiastes 1:9 (NIV) states: "What has been will be again, what has been done will be done again; there is nothing new under the sun." This verse captures a lot of what I'm trying to get across. Man believes he's discovering new things, but those things were always there. It's just that they're being revealed to us as we reach for a higher calling.

A mighty and merciful God designed a system in which man can thrive. But how do we know what God truly wants? The smarter men become, the more they think they are godlike and the more powerful they become. Greater knowledge turns into a source of pride, vanity, and self-worship, which cloud the issues even more when hooked into a political agenda. There's no forgiving of mistakes when it comes to men who fancy themselves as gods or masters of others. Their only intent is to beat down the good that came from our Founding Founders and the notion that we are under God.

Why would God create human beings that exhale 100 times more CO_2 than they inhale? Why command them to be fruitful and multiply? And why is it that the answer to increased CO_2 is plant life? Even if you don't believe in God, the synergism between animals and plants can't be ignored. How do you know if God wants 10 billion people or 20 billion? Only God knows the answer.

This issue is about control by a group that thinks it knows what the planet can and can't hold. God is *used* as a tool now to further the political aim. People are being convinced that they are responsible for destroying our future, *and God would never want that.* Well, guess what? I rather doubt God wants societies of indentured servants to the state either. Nor do I think He wants your values, gratitude, and humility outsourced to some all-powerful human entity.

While Barack Obama's "fundamentally transforming the United States of America" remark is taken out of context (as I understand it, he was trying to convey a message that everyone should have a fair shot), it's morphed into the reality of what we are seeing now — a transformation away from what was designed to give individuals a chance. The very fact that weather and climate are even part of this transformation means that extremists will use any means necessary to get their way.

There's no question that the problems beleaguering this nation and our world today are great, but the advancement of mankind shouldn't be limited by a group that thinks it knows what's best for everyone else. Do we humble ourselves, acknowledge our weaknesses, and perhaps realize that the tests we face are meant to bring us closer to God? Or do we accept putting life, liberty, and the pursuit of happiness on the back burner?

I have offered many examples of how events today are being portrayed as worse than ever, which isn't actually the case. There are three conclusions to draw from this.

1. Shouldn't my being able to pull out so many counterexamples be a source for skepticism toward the prevailing narrative?
2. Why won't alarmists at least acknowledge my side's ideas? Even I admit that CO_2 likely has something to do with global warming (albeit a minor effect compared to other natural drivers).
3. The counterarguments I provided are the tip of the iceberg and expose the obvious strategy of weaponizing the weather.

I always urge people not to believe me. Look for yourself. My goal is to get the headstrong to consider that much of what they believe probably relies on what they don't know. Some may wonder why this is a big deal. Here's why: Climate change may still be ranked low in overall voter importance, but all it takes is 5% of climate-centric voters to tip an election. Hopefully this book will at least make you question what you're being told — and help you choose wisely your course of action.

The crisis is not with climate. It's with the course our nation is taking. And that course pits the foundational values of life, liberty, and the pursuit of happiness against those who wish to limit and destroy those values. The choice is ours. But if the saboteurs win, there may be no going back.

About the author

Joe Bastardi has been in love with the weather since his first memory and is now one of the most experienced operational forecasters in the world with 45 years of experience and no plans to slow down. He is the son of a degreed meteorologist, Matthew Bastardi (Texas A&M University, 1965), and his son Garrett loves the weather and hopes to work in the field someday (once his golf career is over).

Joe has been referred to as a pioneer in extreme weather and long-range forecasting and is often seen and heard on media sites when major storms or climate issues become noteworthy. He is the author of *The Climate Chronicles: Inconvenient Revelations You Won't Hear From Al Gore — and Others*. He graduated from Penn State University (PSU) in 1978, where he wrestled, and is the only known meteorologist since 1950 to letter Division 1 in wrestling. He still applies the work ethic taught to him by his parents and reinforced by his coaches to his job.

At 65, he has no intention of slowing down. He has averaged over 3,000 hours a year of work since starting his first meteorological job in March of 1978. He resides in Boalsburg, Pennsylvania, with his wife Jessica (PSU, 1988), who is a former captain of the PSU women's gymnastics team and has been an assistant coach for 20 years. Their daughter, Jessie, is on the PSU women's gymnastics team.

Above all, Joe has a deep belief in a mighty and merciful Heavenly Father and eternal gratitude for having the chance to do what he was made to do.

CPSIA information can be obtained
at www.ICGtesting.com
Printed in the USA
LVHW022316190221
679378LV00018B/254